42705602

D1511680

Environmental Biology
Laboratory Manual

Alessandra Sagasti

Kendall Hunt
publishing company

Cover image © Andrejs Pidjass, 2009. Used under license from Shutterstock, Inc.

Kendall Hunt
publishing company

www.kendallhunt.com
Send all inquiries to:
4050 Westmark Drive
Dubuque, IA 52004-1840

Copyright ©2009 by Kendall Hunt Publishing Company

ISBN 978-0-7575-7056-8

All rights reserved. No part of this publication may be reproduced,
stored in a retrieval system, or transmitted, in any form or by any means,
electronic, mechanical, photocopying, recording, or otherwise,
without the prior written permission of the copyright owner.

Printed in the United States of America
10 9 8 7 6 5 4 3 2

Table of Contents

E X E R C I S E 1

The Scientific Method

OBJECTIVES

Students who complete this exercise will
- use the scientific method, including writing hypotheses, designing experiments, and conducting experiments to test their hypotheses
- illustrate their results with graphs
- use common experimental design terms, including independent variable, dependent variable, control variable, and replication to describe their experiments

INTRODUCTION

The *scientific method* is a powerful tool for learning about nature. It helps scientists organize and focus their thoughts, and it leads them efficiently to new discoveries. Although the scientific method is useful for studying biology and chemistry, it can also be used in almost any facet your of life when you need to figure something out.

The scientific method involves following a series of steps in sequence. Often, students are surprised that these steps are stated and numbered differently in different science texts. However, although the wording and the numbering might differ, all books outline the same basic steps: making observations, defining a question, stating a hypothesis, testing the hypothesis with an experiment, stating a conclusion, and asking a new question.

In this lab, your lab group will use the scientific method to learn about the behavior of the brown flatworm, *Dugesia tigrina*. The brown flatworm lives at the bottom of ponds, streams, and lakes, where it hunts small invertebrates. It is nocturnal, and it spends much of the daytime underneath rocks and detritus. The brown flatworm can sense light, water movement, and chemicals in its environment.

MATERIALS (PER GROUP)

- 5–10 flatworms
- 3–5 bowls, 250 mL–1L
- 1 ml pipette
- Depending on the details of their experimental design, student groups may need additional materials. Some useful materials to have on hand include pebbles, thermometers, rulers, water baths, lamps, fish food, timers, and other forms of food.

STEP 1—OBSERVATION

Fill a bowl halfway to the top with distilled or deionized water. With a pipette, gently pick up a flatworm and release it into the bowl. Continue adding flatworms until you have 3–5 flatworms in your bowl.

Observe the flatworms for 10 minutes. Make notes on their appearance, movement, and interactions. Notice if they move in any particular direction and how often they change directions. Watch to see how quickly they move, how often their heads turn, whether they swim through the water or glide on the bottom; notice anything else you can observe about them. After watching the flatworms for at least 5 minutes, you can also observe how they respond to stimuli. For example, do they appear to respond to noises, lights, scents, or water movements? Make certain you do not harm the flatworms during your observations.

    1.   Record your observations.

STEP 2—DEFINE THE QUESTION

Based on your observations in Step 1, ask a simple question about flatworms. For example, you could ask a question about the kinds of habitats that attract flatworms. Do your flatworms seem to move toward light or dark environments? Do they move toward flat or complex environments? To simplify your experiment today make sure your question deals with just one variable such as light, temperature, or texture. Also, make sure that your question is as PRECISE as possible.

    2.   What question will you study?

STEP 3—FORM A HYPOTHESIS

Once you have a question, you can begin to form a hypothesis—a reasonable guess about flatworm behavior and how the experiment is likely to turn out. For example, if my question was, "Are flatworms more active in the light or the dark?" I could reasonably guess, based on my knowledge that flatworms are nocturnal, that they might be more active in the dark.

3.   Based on your own knowledge, background information from your instructor, your common sense, and your observations, what do you think the answer will be to your question?

Once you have a preliminary answer to your question, refine it to make it a testable hypothesis. The hypothesis should be as precise as possible. Most importantly, it should be a *prediction* that specifically states what the outcome of your experiment will be. For example, in my flatworm experiment about light levels, I could write the following hypothesis, "If I expose flatworms to varying levels of light, they will spend a greater percentage of time in motion when they are in dark conditions."

4.   What is your hypothesis?

STEP 4—DESIGN AND PERFORM AN EXPERIMENT TO TEST YOUR HYPOTHESIS

There are many possible ways to test any hypothesis—designing your own experiment gives you a chance to be creative. However you decide to carry out the experiment, keep in mind that it should directly address the hypothesis. Also, try to keep it simple! The questions below will help you design your experiment.

Every experiment has an *independent variable*, something the experimenter is interested in studying, and therefore something that he or she varies on purpose. For example, in my experiment I am interested in studying how light affects flatworms. Therefore, the independent variable is the amount of light. I will have two *treatment levels* of my independent variable: a dark treatment and a light treatment. In other words, I will place some flatworms in the dark, and others in the light.

5.   In your experiment, what is the independent variable? What treatment levels of this variable will you investigate?

The *dependent variable* is what you measure. It is critical to know exactly what you will measure before you begin your experiment. For example, in my experiment with light, I want to measure flatworm activity levels. But what exactly does "activity level" mean? I could measure the number of seconds that a flatworm spends moving during several minutes of observation, or I could measure the distance that a flatworm travels in a certain amount of time, or I could measure the average speed of flatworms at set intervals. The best measure is the one that best addresses your question, and that corresponds with your hypothesis. Your measurement should be quantitative.

    6.   What exactly will you measure? How will you measure it?

Next, consider your *control variables*. Control variables are factors that could affect the outcome of your experiment, but that you are not interested in testing at the moment. To keep control variables from influencing the experiment, make sure they are the same in all treatments. For example, light levels may affect flatworm activity, but temperature, texture of the surface, and food availability might also affect flatworm activity. In my experiment, I will control the temperature, texture, and food, so that they are the same in all treatments.

    7.   Name at least 5 control variables in your experiment.

All experiments need *replication*. A *replicate* is the number of times each treatment is repeated. Why do you need to repeat the treatments? Replication helps control for chance events that can affect the experiment. For example, imagine that I perform my light experiment with just 2 flatworms, one in the light and another in the dark. What if the flatworm I chose for the light treatment was an unusually energetic flatworm? I could mistakenly decide that light made flatworms more active. To control for this, I will repeat the light and dark experiments at least several times with different flatworms.

    8.   How many replicates will you use in your experiment?

9. Describe your experiment in as much detail as possible. Make sure you explain exactly what you will do, how many times you will do it, exactly what you will measure, and anything else that a reader might need to know if they wanted to repeat your experiment exactly. You may want to draw a diagram of your experiment.

10. Create a table in which to record your data.

At this point, describe your experiment to your instructor. He or she might have good suggestions for simplifying your experiment. Once your instructor approves the experiment, go ahead and carry it out. Record the data on your data table.

11. On a separate sheet, graph your results. Be sure the axes on your graph are labeled clearly. Keep in mind that the purpose of your graph is to help a reader decide if your hypothesis was supported by the data. Be sure to staple this graph to the rest of the lab.

STEP 5—CONCLUSION

12. What can you conclude from your experiment? Make sure you discuss whether your hypothesis was supported or rejected.

STEP 6—ASK A NEW QUESTION

After each experiment, scientists assess what they learned and decide on the next step. Sometimes they revise their original question or hypothesis. Often, they revise the experimental design if there were problems with the experiment. At other times, the results of one experiment will lead directly to a new, more interesting or relevant question.

13. If you were going to repeat your experiment, how would you improve it? Every experiment can be improved.

14. Based on your results, what new questions would you ask about brown flatworms?

EXERCISE 2

Microscopes

OBJECTIVES

Students who complete this exercise will
- use dissecting and compound microscopes to view living organisms and prepared slides
- calculate total magnification
- make and view wet mount slides

INTRODUCTION

Microscopes are fundamental tools in biology, and have helped scientists make critical discoveries in many biological fields. Early microscopists used microscopes to discover cells. Soon after, biologists used microscopes to view bacteria and other microbes, leading to changes in hygiene in hospitals and homes that greatly decreased human disease. In ecology, microscopes have helped biologists learn about plankton and aquatic food webs, microbial processes in soil, and much more.

Today, you will learn to use two kinds of microscopes: dissecting microscopes and compound microscopes. Learning to use microscopes effectively requires patience and practice. While some students find microscopes frustrating at first, most get the hang of it if they persevere.

As you complete today's lab, take the time to enjoy yourself. Microscopes can show us amazing things! They give us a glimpse into the lives of millions of little creatures around us about which we are barely aware. Have fun!

MATERIALS (PER GROUP)

- Dissecting microscope
- 2 bowls, approximately 500 ml
- Lens paper
- 1 ml pipette
- Live hydra
- Live moss in a dish, barely covered by water for several hours prior to the lab
- Compound microscope
- Microscope slide
- Cover slip
- Prepared slide of a mosquito larva
- Live *Elodea canadensis*
- 1 pair forceps

METHODS

There are a great variety of light microscopes, and although all have much in common, they differ slightly in the location of focus knobs, light switches, and more. Therefore, before you begin, your instructor will point out the main features and parts of the microscopes you will use today.

THE DISSECTING MICROSCOPE

Once you have been introduced to the microscopes, place a small clump of moss in a bowl, and place the bowl on your dissecting microscope.

Turn on the microscope according to your instructor's instructions. Most dissecting microscopes allow you to use light from above or below the object you are viewing. For the moss, use light from above.

Now you are almost ready to look at your moss. First, use a small piece of lens paper to gently wipe the lenses. Place your eyes a couple of centimeters away from the eye pieces, and slowly adjust the eyepieces according to your instructor's instructions until they fit your face. Once you see one single circle of light while using both eyes, you are ready to focus.

Make sure your microscope is on the lowest magnification setting. With your eyes looking through the lenses, slowly turn the focus knobs. Notice that as you turn the focus knobs, the lenses move closer to or farther away from the moss. When the moss is at a distance equal to the focal length of the lens, the image becomes clear and focused. Continue turning the knobs gently until the moss comes into sharp focus.

With the microscope, you should be able to see details you could not see with your eye alone.

1. Are the tiny leaves on the moss smooth or toothed? Are they round or pointed?

2. Draw the moss as viewed through the microscope.

Next, move the magnification to the higher setting and focus again by moving the focus knobs up and down.

3. The entire area you see through the lenses is called the *field of view.* At a higher magnification, are you looking at a smaller field of view or a larger field of view?

4. At a higher magnification, are you looking at a smaller or larger section of the organism?

Moss is often home to small animals, including *tardigrades* (water bears), *nematodes* (roundworms), *oligocheates* (earthworms), and mites. To look for moss-dwelling animals, first remove the moss from its petri dish. You will inspect the debris left at the bottom of the dish. Turn the magnification back to the lowest setting. You may want to switch to the bottom light. With your eyes a couple of centimeters behind the eyepieces, move the focus knob until the image is sharp. Once you see animals, you may increase the magnification and focus again for a closer look.

5. Do you see any small animals moving around your moss? If so, draw a quick sketch of one or two of these animals.

Put aside your moss and the debris with animals. Next, you will look at an aquatic animal, hydra. Hydra lives in fresh water, and is related to marine anemones.

Add distilled water to a new bowl to a depth of 2–3 cm. With a pipette, gently pick up a hydra and place it in your bowl. Place the bowl on your dissecting microscope.

Turn your microscope back to the lowest magnification setting. For viewing the hydra, turn off the top light; instead, turn on the bottom light on your microscope. Place your eyes a few centimeters behind the eyepieces, and slowly move the focus knobs up and down until you see a crisp, clear image of the hydra. Once you have focused at the lowest magnification, you are ready to zoom in some more. Turn the microscope to the highest magnification and, again, turn the focus knobs until you have a clear image.

6. Draw the hydra.

7. Describe one detail about hydra that you can see with the microscope, but not with the naked eye.

8. Why did you use a top light for viewing the moss, but a bottom light for viewing the hydra? If you are not sure, try viewing the moss with the bottom light and the hydra with the top light.

COMPOUND MICROSCOPE

Compound microscopes use multiple lenses (thus, the name compound) to provide even greater magnification than dissecting microscopes.

Examine your compound microscope. Note the *ocular lenses* (the ones by your eyes). Written on the side of each lens is its magnification. Magnification refers to how many times larger than real life the object appears when viewed through that lens.

9. What is the magnification of the ocular lenses on your microscope?

Your microscope will have a number of *objective lenses,* each with a different magnification. Objective lenses are the ones that sit closest to the slide you will observe. To move from one objective lens to another, gently place your hand on all of the lenses and turn slowly until a new one snaps into place.

10. What is the magnification of the smallest objective lens?

11. What is the magnification of the largest objective lens?

When you use a compound microscope, you are always looking through two lenses—the ocular lenses AND one of the objective lenses. Therefore, you are magnifying the object twice—once with the ocular lens and then again with the objective lens. To determine the total magnification when using a compound microscope, multiply the magnifications of the ocular and objective lenses.

12. What is the total magnification when you use the smallest objective lens on this microscope?

13. What is the total magnification when you use the largest objective lens on this microscope?

Before you begin viewing any objects, gently clean all the lenses of your microscope with a piece of lens paper. Then, carefully move the lenses until the smallest objective lens snaps into place. To view any object with the compound microscope, ALWAYS begin with the smallest objective lens. Place a slide of a mosquito larva on the *stage,* the flat area beneath the objective lenses. Most microscopes have clips that hold the slide securely into place. Use these according to your instructor's directions. Practice moving the slide from side to side and up and down using the stage controls.

14. Why is it advantageous to move the slide with the stage controls rather than just moving it with your fingers?

Next, find the *focus knobs* on your compound microscope. Note that there are TWO focus knobs, a large one (the *coarse focus*) and a small one (the *fine focus*). Gently move each focus knob and note the movement of the stage.

15. How does the movement of the stage differ when you move the coarse focus compared to using the fine focus?

When using a compound microscope, use the coarse focus with the lowest magnification objective lens. However, when you move to higher magnifications, ONLY use the fine focus, or you can damage the lenses.

16. Why would using the course focus with a high magnification lens possibly damage it? Hint: Compare the length of the objective lenses.

Next, turn the microscope on and adjust the ocular lenses to fit your face following your instructor's directions. Use the stage controls to slide the mosquito larva until it sits above the light. With your eyes a couple of centimeters behind the ocular lenses, gently move the coarse focus knob until the edges of the image are crisp.

Find the *iris* on your microscope, and gently move it from side to side while viewing your mosquito.

17. What does the iris do?

Many microscopes also allow you to adjust the brightness of the light source. If your microscope allows this, try adjusting the brightness while viewing your mosquito.

The iris and brightness adjustment can help you work more comfortably and effectively (many students ignore these and end up with headaches from using excessive light). Most students find it useful to adjust the iris whenever they view a new object or change to a new objective lens.

Make sure the mosquito larva is still focused crisply at the lowest magnification. Next, gently slide the next smallest objective lens into place and look through the eye pieces. Turn the fine focus knob until the image is sharp. If the microscope was focused carefully at the previous magnification, you should not have to move the focus knob very far. Adjust the light using your iris or brightness adjustment. If you want, you can now move to the next highest magnification, focus with the fine focus knob, and adjust your light.

18. Find and draw your mosquito larva's head.

19. What structures on its head does the mosquito use to filter food from the water in which it lives?

On the rear end of your mosquito larva is a long tube called a siphon. Mosquito larvae place their siphons at the surface of the water to reach oxygen in the air above their ponds.

20. Draw your mosquito larva's siphon.

21. How does the siphon help the mosquito larva to survive in stagnant water?

Another common inhabitant of ponds is the American waterweed, also known by its scientific name, *Elodea canadensis.*

22.   Draw *Elodea canadensis* as you see it with your naked eye.

To make a slide of *Elodea canadensis,* gently use forceps to pull one leaf off a plant. Place the leaf in the center of a glass slide. Add one drop of water, and then place a cover slip on top of the leaf and water. Place your slide on the stage, securing it with clips. Using the stage controls, move the slide until the leaf is directly above the light.

Make sure the microscope is set with the smallest lens facing the object. While looking through the eye pieces, turn the coarse focus until the plant is sharp. Adjust the light. Move to the next lowest magnification lens, and adjust the focus using the fine focus knob. Adjust the light. Finally, move to the lens with the highest magnification, focus with the fine focus knob, and adjust the light. You should be able to view individual cells at this magnification.

23.   What is the total magnification at which you are viewing this plant?

24.   Draw a couple of *Elodea canadensis* cells, showing details of the outsides and insides of the cells.

25.   The cells are surrounded by thick cell walls. What do you think is the purpose of these walls?

Inside *Elodea canadensis* cells you will see small balls called *chloroplasts* that often move around within the *cytosol,* the jelly-like contents of the cell.

26. Plants *photosynthesize,* or make food, using a green pigment called *chlorophyll.* Given this information, where inside the cell does photosynthesis take place?

27. Describe at least one detail about *Elodea canadensis* that you could only see when using a microscope.

DISCUSSION QUESTIONS

28. Dissecting and compound microscopes are both useful to biologists, but they are used in different situations. Describe a situation in which using a dissecting microscope would be most appropriate, and explain why it would be more appropriate than a compound microscope.

29. Describe a situation in which a compound microscope would be most appropriate, and explain why it would be more appropriate than a dissecting microscope.

30. What did you see today that surprised you? What else do you think would be interesting to view with a microscope?

EXERCISE 3

Dichotomous Keys

OBJECTIVES

Students who complete this exercise will
- use dichotomous keys to identify unknown specimens
- create dichotomous keys

INTRODUCTION

Scientists have described and named more than 1.8 million species. Millions of additional species remain undescribed and unnamed. With so many species, it can be difficult to identify a particular tree, bird, worm, or snail when you find one. When scientists come across organisms they don't recognize, they often turn to *dichotomous keys*. Dichotomous keys guide you quickly and efficiently to a correct identification by helping you notice the organism's most distinguishing characteristics. They can also be lots of fun, making a hike or a walk in the woods far more interesting because you can identify what you see.

Dichotomous keys offer you a series of choices to help you zero in on the correct identification. For each choice, there are usually two mutually exclusive descriptions, one of which applies to the unknown specimen. When you reject the inappropriate description, you eliminate from consideration any species whose characteristics don't match the unknown specimen. Each time you exclude some possible identifications, you close in on the correct identification.

To identify an unknown specimen using a dichotomous key, compare its characteristics to those described on the key. Start at the first couplet, or series of two choices, and select the description that best applies to your specimen. When you determine the best choice, the key will direct you to the next couplet that will again ask you to choose between two descriptions. Always read both choices offered in each couplet (sometimes the first choice seems right but when you read the second you realize that it applies more clearly to your specimen). Eventually, you will come to a definitive answer, or positive identification.

The best dichotomous keys use characteristics that minimize mistakes. These keys are unambiguous. For example, rather than asking you if your specimen is small or large, words with different meanings for different people, they will ask you if your specimen is longer than 10 cm or shorter than 10 cm. Good dichotomous keys also focus on obvious traits rather than obscure traits, and avoid traits that vary within a species. For example, in some marine snails, color varies greatly among individuals in the same species. Therefore, color would not be a useful trait for a dichotomous key of these marine snails.

Dichotomous keys do have limitations. Some include only a subset of the species in your area. Therefore, the specimen you want to identify may not be included in the key. This is especially true for botanical keys, which often lack the numerous exotics that we import and plant in our yards. Some animal keys can identify adults, but will not help you identify larva. Other keys will only work during certain seasons. For example, many dichotomous keys to trees are useful only during the growing season, when you can examine leaves on deciduous trees. Finally, sometimes scientists need to identify specimens that were damaged during collection. For these specimens, dichotomous keys may not work, and the researcher will have to ask an expert for help.

Today, a series of activities will familiarize you with dichotomous keys. Specifically, you will
  • use a dichotomous key to identify common mushrooms from a grocery store
  • work with the entire class to create a key to the students in the class
  • create a dichotomous key for identifying seashells, insects, plants, or other local organisms
  • use a dichotomous key to identify campus trees

MATERIALS (PER GROUP)

  • A variety of fresh mushrooms from this list: portabella, white button, shiitake, oyster
  • 10 specimens of different species, all from same taxonomic group, from the local area (for example, 10 local seashells, flowers, insects, or other appropriate group)
  • dichotomous key to common trees in the area (the National Park Service and the National Arbor Day Foundation, among others, sell inexpensive and excellent keys)

MATERIALS (PER CLASS)

  • pictures, pressings, or fresh samples from trees with each of the following characteristics: opposite leaves, alternate leaves, simple leaves, compound leaves, toothed margins, and entire margins

PART I. DICHOTOMOUS KEY TO COMMON GROCERY STORE MUSHROOMS

Use the following key to identify your mushrooms.

1a. Gills (small flaps underneath the cap) are brown
or are not visible ....Go to number 2
1b. Gills are white or off-white ....Go to number 3

2a. Cap diameter is greater than 7 cm ....portabella
2b. Cap diameter is less than or equal to 7 cm ....white button

3a. Stem is on one edge of a smooth cap, and gills from cap
descend onto the stem ....oyster
3b. Cap is rough and brown, often with darker-colored area
directly above stem, stem not on edge ....shiitake

1. Were you able to identify all of the mushrooms? If not, which ones gave you trouble? Why?

2. Were any couplets ambiguous or confusing? If so, which ones? Why were they confusing?

3. Describe another characteristic that could distinguish between these mushrooms. Make sure your characteristic is unambiguous.

PART II. DICHOTOMOUS KEY TO THE STUDENTS IN THE COURSE

The class will work together to create a dichotomous key for identifying all students in the class by name. The first step to writing a dichotomous key is to draw a flow chart organizing the organisms by their characteristics. For example, for the mushrooms above, we could have drawn the following flow chart.

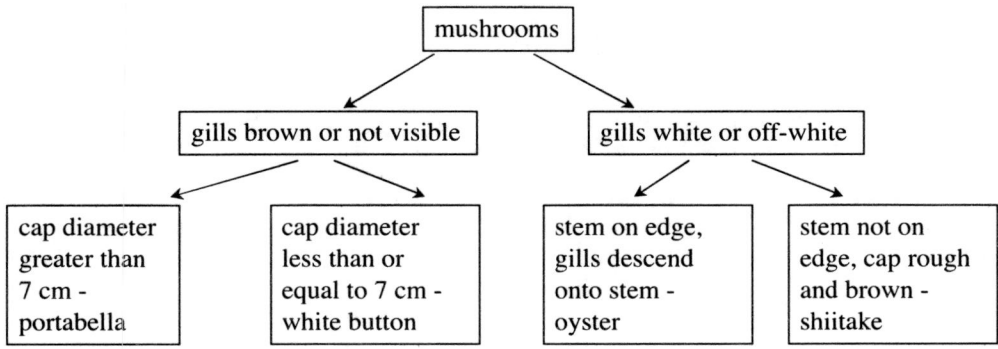

Once you have a flow chart, you can then write out the key in words.

To create a class flowchart, all students should stand in one group. Decide as a class on an easily observable characteristic that separates the class into two groups. Students should physically move and stand in two separate groups. Next, divide each of those groups into two groups, and continue until each group contains just one student. As the class develops the flow chart, the instructor will draw it on the board.

Keep the following rules in mind while developing a flow chart.
- Make sure there are only two choices at every step.
- With each couplet, divide the remaining students into roughly equal groups rather than separating one student at a time. This will help users arrive at identifications more quickly.
- Avoid easily misunderstood terms, such as "tall" or "short." Use precise terms, such as "taller than 6 feet" or "shorter than 6 feet."
- Respect your classmates and avoid suggesting characteristics that could hurt feelings, such as weight.
- Consider using gender, height, eye color, hair color and texture, presence or absence of earrings, and other easily observable characters.

The instructor may choose to invite a visitor who does not know the students into the classroom, and ask him or her to identify students by name using the class dichotomous key. Using the visitor's advice, the class can discuss whether the key needs revision.

4. On a separate sheet of paper, draw the class flow chart. Your instructor may also ask you to write a dichotomous key in words using the flow chart as a guide. Be sure to attach these to your lab.

5. Which of the couplets do you think was particularly effective? Why?

6. Did any couplets need revision? If so, why?

7. Would the class key work in future lab periods? Will it work during all seasons of the year? Will it work in several years when your classmates are older?

PART III. CREATE A DICHOTOMOUS KEY FOR LOCAL SPECIES

With your group, observe the assigned organisms. Create a flow chart dividing the species into groups based on easily observable characteristics. Finally, use your flow chart as a guide for writing a dichotomous key.

8. On a separate sheet of paper, draw the flow chart for identifying these local species. Remember to attach it to your lab.

9.   On a separate sheet of paper, write a dichotomous key in words using your flow chart as a guide. Remember to attach it to your lab.

10. Ask your instructor or a student from another group to use your key to identify specimens, and suggest improvements to your key. What improvements did the instructor or student suggest?

PART IV. USE A DICHOTOMOUS KEY TO IDENTIFY LOCAL TREES

Your instructor will show you pictures or examples of trees with the following characteristics. For each one, draw yourself a quick sketch to help you remember the terms. Your instructor may also choose to teach you additional tree terms.

Alternate leaf arrangement   vs.   opposite leaf arrangement

Simple leaf   vs.   compound leaf
(make sure your drawing clearly shows the bud(s))

Entire   margin   vs.   toothed margin

The instructor will select trees on campus or near campus for you to identify using a dichotomous key. With your group, identify each tree. If you don't know some of the terminology on your key, look for a glossary, or discuss these terms with your instructor.

11. Identify each tree selected by your instructor. As you work through the key, write down each choice you make along the way. That way, if you make a mistake, you can go back and see where the mistake occurred.

1.

2.

3.

4.

5.

ADDITIONAL QUESTIONS

12. In addition to dichotomous keys, scientists can use guidebooks to identify unknown organisms. With guidebooks, readers simply compare unknown specimens to drawings or photographs until they find one that matches.

    a. What are the disadvantages of guidebooks compared to dichotomous keys?

    b. What are the advantages of guidebooks compared to dichotomous keys?

Chemical Defense in Plants

OBJECTIVES

Students who complete this exercise will
- determine if herbal extracts deter ant feeding
- use the scientific method
- design and conduct experiments to test hypotheses about chemical defense
- illustrate their results with graphs

INTRODUCTION / OBSERVATIONS

Many prey organisms have evolved adaptations to protect themselves from predators. For example, some blend in with their environment so that predators can't spot them. Others cover themselves with hard shells. Others run away quickly.

Plants also need to protect themselves from *grazers,* organisms that browse on plants. Some plants protect themselves with thorns. Others have fuzzy, hairy leaves that are unpleasant for grazers to eat. Many plants have very high growth rates, so when they lose some of their leaves they can quickly replace the lost leaves with new ones. And still other plants use *secondary chemicals.*

Secondary chemicals are molecules that plants do not need for primary metabolism. In other words, plants don't use these chemicals for photosynthesis, reproduction, or growth. So why do they make secondary chemicals? Many secondary chemicals appear to be unpalatable (bad-tasting or even poisonous) to insect grazers such as caterpillars and grubs. By producing these chemicals, plants may decrease grazing by insects.

Secondary chemicals often give plants distinctive scents and flavors. For example, if you've walked through a pine forest, you've probably smelled pine resin, a sap rich in secondary chemicals that deter potential pine grazers. You may also be familiar with secondary chemicals in your food. Since they have unique flavors and scents, people often use plants rich in secondary chemicals as herbs and spices in the kitchen.

Today, you will work with ants, a ubiquitous insect in a variety of ecosystems. Most ants are not direct plant grazers but instead are decomposers and scavengers. However, many chemicals that are distasteful to true plant grazers such as caterpillars and grasshoppers are also distasteful to ants.

In this lab, you will offer ants food surrounded by extracts of various plants, to try to answer the question: "Which plants are protected by secondary chemicals that deter insects?"

With your lab partner or group, inspect the plants available for this experiment. You may want to feel, smell, and/or do literature searches about these plants. Do NOT taste the plants (although taste is an important variable, tasting substances in a laboratory is unwise because they could be contaminated with other laboratory chemicals).

1. Based on your observations, which (if any) of these plants do you predict will contain secondary chemicals that deter ants? Why?

HYPOTHESIS

2. Write a hypothesis about which plants (if any) will contain chemicals that discourage ant feeding. Make sure you hypothesis is a prediction. You may choose to simplify your hypothesis so that it deals only with a couple of fresh plants rather than test all the choices available in lab today.

MATERIALS (PER GROUP)
- A selection of fresh herbs, including at least some of the following: lavender, mint, sage, and catnip
- 1 mortar and pestle
- 1 balance
- 10 ml graduated cylinder
- Fruit jam or jelly
- Toothpicks
- 10–15 pieces of string or twine approximately 7 cm long
- 1 labeling marker
- 10–15 microscope slides

METHODS

Ant feeding and chemical defense can be studied using the following methods, *which you will need to modify to test your own hypothesis.*

OBSERVING ANT FEEDING

- Locate an anthill on the campus grounds. The ants do not need to be any particular species, but you should avoid fire ants, other ants that sting, and very large ants.
- Using the tip of a toothpick, place a small dab of jam or jelly in the center of a microscope slide.
- Place the microscope slide no further than 50 cm from the anthill entrance. Make sure the slide touches the ground on all sides. If you have to, move leaves and other objects out of the way or use your hands to flatten the ground around your slide.
- Every 5 minutes for 1 hour, record the number of ants actively feeding on the jam.

PREPARING AND PLACING PLANT EXTRACTS FROM A FRESH PLANT

- Weigh 1 gram of plant material, taking care to select tender leaves rather than hard stems, and place in a mortar.
- Add 10 ml of water to the mortar. Using the pestle, crush and grind the leaves for at least 5 minutes to create a concentrated extract.
- Dip a string in the extract until it is thoroughly wet. Place the string in a ring on the surface of the slide around the jam as shown below.

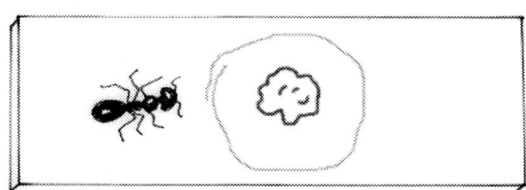

You and your group will modify these basic methods to test your hypothesis. Answer the questions below to help you plan your study.

3. What is your *independent variable?* (What will you vary on purpose?)

4. What are the treatment levels of your independent variable that you will test?

5. At least one of your treatment levels for this experiment should be a *control treatment.* What will you use for a control treatment, and why do you need it?

6. You will also have many *control variables*. Name several control variables in your experiment.

7. What is your *dependent variable?* (What exactly will you measure?) Consider how to decide if an ant is feeding or not. Will you count all ants on the slide? Only those touching the jam? Or will you use some other criteria?

8. How many *replicates* will you have? Explain what your replicates will be.

This is a good time to think about how you will deploy your replicates. The placement of samples around the anthills could determine how quickly ants find your samples, and therefore could drastically alter the results. There are many possible ways to deal with this problem:

*   You could start timing data for a slide only after ants discover it.
*   You could use a *grid* and a *list of random numbers* to randomly place samples around the anthills. Your instructor may choose to discuss this option with the class.
*   You could purposefully place the samples in such a way that all levels of the independent variable have an equal chance of being found quickly by ants.
*   You could devise another method of randomly placing objects around anthills.

9. How will you deploy the replicates to minimize the effect of sample placement on your results?

10. Briefly explain how you will conduct your experiment. Make sure you give enough details that a reader could repeat your experiment. If necessary, draw a simple diagram to explain your experiment.

11. On a separate sheet of paper, draw a table for recording your data.

Discuss the experiment with the instructor before you continue. Once the instructor approves your plans, go ahead and begin your experiment.

RESULTS

Record the results on your data table. Then, answer the questions below.

12. On a sheet of graph paper, plot a graph to depict your results. Make sure you place the dependent variable on the y-axis (the vertical line). On the x-axis, plot time since the start of your experiment. You will need to use different colors or types of lines for your independent variables. Make sure you label the axes and put an appropriate scale on your axes (remember, you want the data to take up most of space on the graph).

DISCUSSION

13. Do the results support your hypothesis, or cause you to reject it? Refer to the data to back up your answer.

14. Were there any problems with the experiment? If so, what were they?

15. Can your results today be generalized to apply to all ant species? To all seasons? All weather conditions? Why or why not?

16. Can your results with the ants be applied to other organisms? For example, if the ants are
     deterred by basil, will basil also deter important agricultural pests? If so, why? If not, why
     not?

17. Why might some plants make compounds that *attract* ants?

18. Other organisms besides plants use chemical defenses. Name several that might be found
     in the areas around campus.

19. How could you improve this experiment? Every experiment can be improved.

E X E R C I S E 5

Succession on Fallen Logs

OBJECTIVES

Students who complete this exercise will
- describe the species composition and species richness of communities on fallen logs
- use the scientific method
- design and conduct experiments to test hypotheses about succession and diversity on fallen logs
- illustrate their results with graphs

INTRODUCTION

Ecological *succession* is an orderly sequence of events that occurs in a community after a disturbance or following the opening of new habitat. In succession, the types of organisms in a community vary predictably over time. For example, when glaciers retreat they leave behind barren landscapes with few inhabitants. Lichens and mosses colonize these areas first, followed by small weeds and shrubs, and finally by large trees. This sequential change in *species composition*, the type of species present in a particular area, characterizes succession.

In addition to changes in species composition, diversity also varies during succession. Often, the earliest stages of succession have low diversity because few species can survive the challenging conditions found immediately after many disturbances. Later, diversity rises and falls in complex patterns as habitat conditions change and as interactions between organisms intensify.

Most forests are littered with numerous fallen logs. Today you will investigate the events that occur after trees fall, to determine if fallen logs lead to succession. As they decay, fallen logs commonly provide habitat for fungi, snails, insects, lichens, mosses, tree seedlings, and more (see Table 1). You will determine if these organisms are equally likely to be found on freshly fallen logs and on logs that are in an advanced state of decay, or if some are found only in logs undergoing particular stages of decay.

While it can be difficult to determine exactly how long ago a particular log fell, we can make a rough approximation of its relative age by looking at the extent of decay. Scientists use visible changes in fallen logs to put them into groups called *decay classes*. For today's exercise, you will put logs into the following decay classes:

- Class 1 logs fell recently. Their bark is securely attached to the trunk, and often there are small twigs or leaves visible on the tree.

- Class 2 logs fell longer ago. In these trees, the bark is loose or partly missing.
- Class 3 logs fell even longer ago. No bark remains, but the trunk material is still hard and solid.
- Class 4 logs are old. They have no bark. The trunk is soft and squishy, and may even be falling apart in places.

OBSERVATIONS

With your instructor, observe a fallen log. Look carefully at the organisms living on top of the log. The instructor will point out lichens, mosses, insects, and more, if they are present. Note if the organisms are living on top of bark, underneath bark, or in the trunk itself. Note any other observations that that you think are important.

HYPOTHESIS

Your group will determine the *species composition* (which species are living on each log) and *species richness* (the number of species living on each log) of logs in each decay class.

If succession occurs in this community, then the species composition will change as logs decay. In other words, you will find some organisms on young logs, a different set of organisms on older logs, and a different set of organisms on the oldest logs.

1. Refer to the list of organisms in Table 1. Using the information provided in the table, library research, and your prior observations of nature, select ONE of the organisms on the list and write a hypothesis predicting in which log class you expect this organisms to be most commonly found, if any.

2. Explain why you made this prediction.

3. Will you find a relationship between decay class and species richness? Write a clear and specific hypothesis that predicts the outcome of your experiment.

MATERIALS (PER GROUP)

- 1 meter stick or a piece of string that is 1 m long
- 2 pieces of string 20 cm long, marked in the center with a marker
- 2–3 hand lenses
- 1 field microscope, or several jars in which to put specimens before taking them back to the lab to examine in dissecting microscopes (optional)
- 1 dissecting microscope (optional)
- Local identification guides for fungi, mosses, and insects (optional)
- 1 clipboard
- 1 pencil or pen
- Access to a forest with many fallen logs

METHODS

Communities on fallen logs can be studied using the following basic methods. You will need to take these basic methods, make them more detailed, and discuss your plans with your instructor before you begin your study.

FINDING AND SELECTING LOGS TO STUDY

For this experiment, your group will select logs to study from among all the fallen logs present in the area your instructor has chosen. To prevent bias from affecting your study, decide ahead of time how to choose from among these logs. Your instructor may provide more guidance about this. As you consider how to select logs, remember that you need to examine logs in a variety of decay classes. You should also select logs that are representative of all the logs in the forest. In other words, if you select only logs that are immediately next to a walking trail, you might get different results from studying logs that are away from the trail. You might also get skewed results if you only choose logs next to streams, or only logs from a particular species.

You may need to restrict the logs you choose for practical reasons. For example, you should probably exclude logs harboring poison ivy or logs that are found in steep or dangerous areas. Keep these restrictions in mind when you interpret the data. Your results will only apply to those types of logs that you were able to include in your study.

4. How will you select logs for your study? Write a detailed explanation.

5. Will you exclude any logs on the basis of location, size, or for any other reason? Explain why you will exclude these logs.

6. How many logs of each decay class will you study? In other words, how many replicates will you have?

7. If you can't find enough logs in a particular decay class, what will you do? (Field studies often run into these types of problems, and it's helpful to have a plan ahead of time for how you will handle the problem if it arises.)

SELECTING A SPOT ON THE LOG TO STUDY

Once your group selects a log, classify it by decay class and choose a spot on the log to study in detail.

8. How will you classify the log if parts of it appear to belong to one decay class, and other parts to another decay class?

9. Is it good science to pick the most interesting part of the log (with the most organisms or the most unique organisms) to study? Or do you want to choose a representative spot? Why?

Fallen logs have many *microhabitats*, areas with unique *abiotic conditions* that harbor different communities. For example, the abiotic conditions and the species found underneath logs can differ from those on top of logs. Similarly, the habitat on the interior of fallen logs differs greatly from the habitat on the edges of logs. To simplify your study, limit your samples to the top surfaces of logs.

CHARACTERIZING THE COMMUNITY

Once you've chosen a log and a spot on the log to study, place your meter stick or string along the top of the log. You will characterize the community of organisms found on a section that is 1m long (the length of your string or stick) and 10 cm on each side of this meter. Use your 20 cm string pieces to determine the area as shown below.

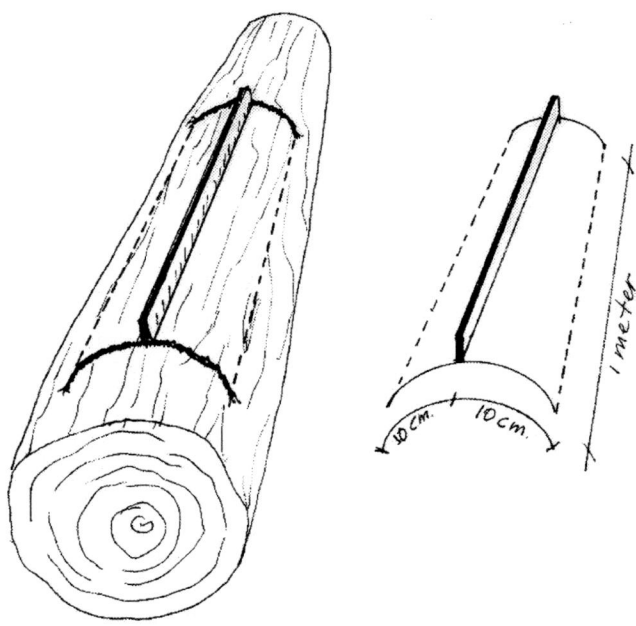

Using a hand lens, examine the area and note all visible organisms. Look carefully, because some of the species are small and inconspicuous. Remind yourself about what you are searching for by looking at Table 1. As you examine the area, you want to determine 1) the number of different species present and 2) the types of species present. It is not necessary to identify every organism. But you should put all organisms into basic categories, such as fungi, lichen, moss, etc. If you find more than one type of fungus, lichen, or other organism, and you cannot identify them to species, simply call them fungus #1, fungus #2, etc., and write yourself a quick description to remind yourself what each one looks like. If you get stuck and cannot place an organism into a group, put a small sample of the organism in a jar, label it, and identify it later, with your instructor's help, using a dissecting microscope and identification guides.

10. On a separate sheet of paper, make a table for recording your data. Make sure you leave room on this table for recording each replicate of each decay class, and leave yourself adequate space for taking notes.

RESULTS

11. List the organisms found in each decay class.

    • Class 1

    • Class 2

    • Class 3

    • Class 4

12. On a separate sheet of paper, draw a graph illustrating the relationship between decay class and species richness. Remember to label your axes.

DISCUSSION

13. What is succession?

14. Which (if any) organisms appear to be most common in the early decay classes? Explain by referring to your data. Why do you think these organisms are common on logs that are just beginning to decay?

15. Which (if any) organisms appear to be most common in the later decay classes? Explain by referring to your data. Why do you think these organisms are common on logs in late stages of decay?

16. Look back at your hypothesis for question 1. Was your hypothesis supported or rejected? Explain by referring to your data.

17. Is there clear evidence of succession in your data? Why or why not?

18. What happened to diversity (species richness) as the logs decayed? Was your hypothesis about species richness (question 2) supported or rejected? Why do you think you saw this pattern (or lack of pattern)?

19. Which of the organisms you saw are detritivores? Which are decomposers?

20. How do detritivores and decomposers influence the forest community?

21. In many woodlots, when trees fall down, they are removed and sold as lumber. How does this practice affect the forest? Think carefully and explain at least two results of this practice.

22. How would you change your experiment if you were to repeat it? Every experiment can be improved.

Table 1. Organisms commonly found on fallen logs.

Type of organism	Appearance	Role in community	Special notes
Fungus	Fungi are made of thin strings that can be hard to see. But fungi also produce mushrooms or other large reproductive structures that are easier to see. These are usually soft, often have gills, and can be brightly colored.	Decomposers or parasites	Look carefully – many fungi will show up only as tiny specks of color on the trunk.
Moss	Small, green fuzzy plants that are never more than a few centimeters tall.	Producers	There are many species of moss. Look carefully at the shape of their leaves to determine if you have more than one growing in your area.
Lichen	These are hard, crusty structures made by fungi and algae that live together in mutualistic relationships.	The algae in lichen are producers.	There are many species of lichen. Look carefully at the color and shape to see if you have more than one.
Insects	Ants, beetles, termites, and other insects are common on logs. All have hard exoskeletons and segmented bodies.	Grazers, detritivores, and predators.	You can decide with your group if you want to count holes, tunnels, or other signs left by insects.
Spiders	Spiders have 8 legs and often build webs.	Predators	If you see a spider web, but not a spider, you can be sure the spider is hiding nearby. Therefore, you can count this spider as a sample.
Tree seedlings	These look like typical, but young, plants.	Plants are producers.	There may be more than one species. You can tell species apart by looking at the shape of the leaves.

E X E R C I S E   6

Using Bar Graphs to Examine Student Transportation Habits

OBJECTIVES

Students who complete this exercise will
- conduct surveys of student transportation habits
- plot three graphs to illustrate the survey results, including bar graphs with one independent variable, two independent variables, and with error bars
- use graphs to analyze data and evaluate hypotheses

INTRODUCTION AND OBSERVATIONS

Graphing is an essential tool in science and in many other fields. Graphs help organize and analyze data. Scientists also use graphs to communicate complex information clearly to other scientists and to the general public. Graphs can be used to persuade as well as to inform. Because graphs can be misused and manipulated to mislead others, it is important to examine graphs critically.

While most students can draw and interpret simple graphs, almost everybody can learn to make their graphs more effective. Common questions that even seasoned pros have about graphing include the following: How can you use error bars to convey information about variability in your data set? How can you draw graphs to avoid misleading readers? How can you make your graphs as appealing and informative as possible?

During today's laboratory session you will survey students on your campus about their transportation choices, and use graphs to analyze the results. Transportation choices help determine the types and amounts of resources used by society, and greatly influence relationships between nations. Air pollution, water pollution, traffic, and life styles are all greatly affected by transportation choices. In addition to asking students about their transportation choices, you will also record their gender, where they live, and their student status to determine if there are relationships between these factors and how students choose to get from place to place.

HYPOTHESES

In this survey, you will ask students on your campus how they get to school. Examine the "Transportation Survey" at the end of this lab, and then make hypotheses about the outcomes.

1. Which transportation option (driving alone, carpooling, public transport, walking, biking, or other) will students at your school use most often? Explain how you made your prediction.

You will also consider whether gender, student status (full or part-time student) or distance between a student's home and school affect transportation choices.

2. Will the survey show a relationship between transportation choices and gender, student status, or distance from school (pick one of those three)? If you think there will be a relationship, what will it be? If you don't expect to see a relationship, why not?

Finally, you will ask students how often they use public transportation.

3. Will the survey show a relationship between public transportation use and gender, student status, or distance from school (pick one of those three)? If you think there will be a relationship, what will it be? If you don't expect to see a relationship, why not?

MATERIALS
- At least 3 sheets of graph paper per student
- 5–10 copies of "Transportation Survey" per student
- 1 pen and clipboard per student
- 1 computer shared by the class, with a spreadsheet program

METHODS

You and your classmates will spread out around campus to survey students. Consider surveying students in the cafeteria, library, outdoor areas, study lounges, and hallways. Each class member should ask between five and ten students (you instructor will decide the exact number) to answer the survey. Once you've collected enough student responses, return to the laboratory to pool your results with those of your classmates. When the class data set is complete, use graphs to analyze the data so you can evaluate your hypotheses.

To ensure that survey results accurately portray the transportation choices of students on your campus, the class needs to avoid *sampling bias*. Sampling bias occurs when samples are non-representative, leading to results that don't accurately reflect the population being studied. For example, if you conducted all the surveys at a bus stop, students who take public transportation would probably be overrepresented, and not reflect the student body as a whole. If any individual students fill out more than one survey, they can also skew the results. To avoid this, ask all potential respondents if they have already filled out the survey. If so, halt the survey and move on to survey other students. Bias may also occur if students don't answer the survey truthfully. To minimize this problem, don't comment as respondents fill out surveys.

As a class, discuss a plan for surveying students that will minimize sampling bias. The class may choose to assign students to a variety of locations on campus or to sample specific groups of students.

4. Where will you take your samples? Will you target a specific group of students?

Once the class and the instructor agree on a plan, go ahead and gather the data. When you return to the lab with completed surveys, enter the data into a spreadsheet set up by your instructor. Once all of your classmates have entered their data, the instructor will sort, average, and manipulate the data as needed and place the information on the board or in another easily accessible location. Use these data to graph the class results and evaluate your hypotheses.

Although your instructor will use an electronic spreadsheet to compile and sort the data, you should draw all graphs for this laboratory exercise by hand. Although computers can help you draw excellent graphs once you have good graphing skills, they can also distract you from learning the basics of good graphing. Rather than learning how to graph on any particular computer program, first practice constructing graphs on your own. Then you'll be able to use graphing programs more effectively in the future.

RESULTS

As you begin graphing, remember these rules of thumb:
- Before drawing any graph, make sure you can state exactly what information you want the graph to convey.
- Make sure graphs depict the information a reader will need to evaluate your hypothesis.
- In general, the horizontal axis, known as the x-axis, should represent the independent variable.
- In general, the vertical axis, known as the y-axis, should represent the dependent variable.
- The bars, lines, or dots on your graph should take up most of the graph's space. If they don't, adjust the axes to make better use of space.
- Always label the axes, and include units.
- Give your graph a title.
- Include a legend when necessary.
- Usually, simple graphs are clearer than complicated ones. Therefore, draw graphs as simply as possible. If appropriate, graph averages rather than all of the data (but also include error bars or another measure of variability).

BAR GRAPHS

Start by drawing a simple bar graph. Bar graphs can be particularly effective when comparing proportions. For example, the following graph compares the transportation modes of American workers over 16 years old, as reported in the 2000 census.*

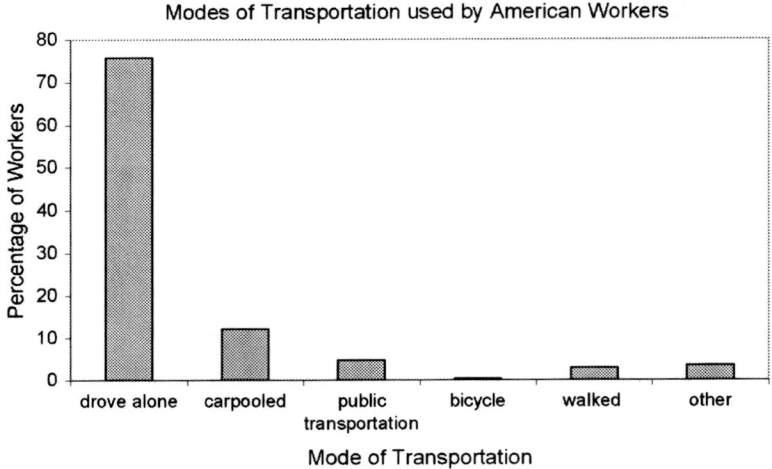

5. What type of transportation was most used by American workers in 2000? Why do you think this was the case?

6. What type of transportation was least used by American workers in 2000? Why do you think this was the case?

7. On a separate piece of graph paper, draw a bar graph showing the transportation choices of students at your school.

* data from: Reschovsky, Clara. "Journey to Work: 2000." March 2004. U.S. Department of Commerce, Economics and Statistics Administration, US Census Bureau.

USING BAR GRAPHS TO COMPARE TWO POPULATIONS

To compare data from two populations, you can place two bar graphs side by side, or plot two data sets on a single graph. When plotting multiple data sets on one graph, always use different colors or patterns to differentiate the data sets and include a legend, as shown in the graph below.

Weight Loss Experiment

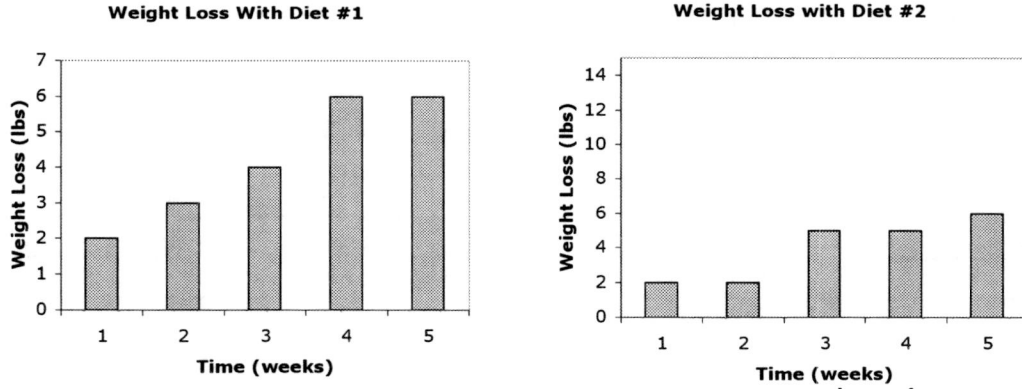

If you place two graphs side by side, be sure to use the same scale on both graphs to avoid confusion. In fact, unethical graph-makers can manipulate scales to fool viewers who don't watch out for this trick. For example, the two graphs below show the weight loss from two different diets, but are difficult to compare because they have different scales.

8. According to the graphs, which diet caused the greatest weight loss by week 5? Why were the scales on the two side-by-side graphs potentially misleading?

9. Review the hypothesis you wrote for question 2. Then, on a separate sheet of graph paper, draw a bar graph to depict the transportation habits of students with different genders, student status, or distance from school, depending on your hypothesis. Use different colors or patterns to represent males vs. females, part-time students vs. full-time students, or students that live less than 5 miles, between 5 and 10 miles, or more than 10 miles from school. Include a legend.

ERROR BARS

Often, it's easier to analyze graphs that show averages rather than all the data points individually. Graphs of averages are simpler, and therefore often easier to interpret. However, when you graph averages, you also need to provide viewers with some idea of the variability (spread) in the data. For example, imagine a researcher was interested in determining the number of hours that male and female students at a university studied for a particular test. Her hypothesis is, "Females spend more time studying than males." Let's say this researcher polls her friends and gets the following data:

Hours Spent Studying	
Males	Females
2	1.5
4	4
5	6
6	8
4.5	4
3	3
5.5	2
7	2
1	1
4	3.5

The graphs below depict the same data in three different ways.

Study Habits of Male and Female Students

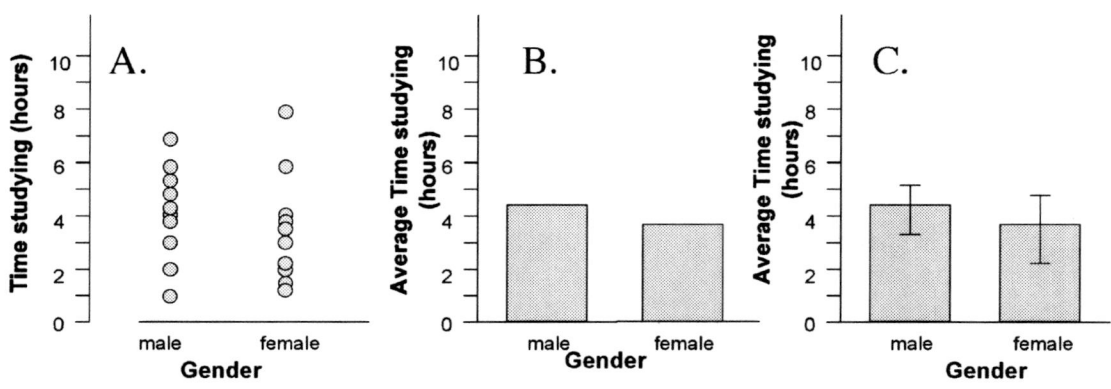

- Graph A gives a good indication of the spread (also known as *variability*) of the data, but it's difficult to determine which gender studied more.

- Graph B shows averages, and appears to show that males spend more time studying than females. However, this graph doesn't show variability. Since the researcher didn't study every single student at the university, her estimate using a sample of her friends probably doesn't exactly reflect what happened in the entire student body. But how good is her estimate? With no indication of variability, we can't tell.

- Graph C shows averages, and uses error bars to indicate variability. In general, the longer the error bar, the more variability. This particular type of error bar is a statistic called the *95% confidence interval.* That means that there is a 95% chance that the true average for all males at the university is somewhere within the error bar. The same is true for females – there is a 95% chance that the true average is somewhere within the error bar for females. Notice that the male and female error bars overlap substantially. In other words, we cannot tell which average is truly higher. In the end, the researcher has to reject her hypothesis. Females do not spend more time studying.

There are many ways to calculate error bars, and interesting statistical theory to explain these calculations. If you take a statistics course, you will learn much more about confidence intervals and error bars. For now, you will learn one simple way to calculate 95% confidence intervals for most data sets.

For this class, we will calculate 95% confidence intervals as follows.
- Most spreadsheet programs can calculate a statistic called a *standard deviation.* Your instructor will show you how to do this and calculate it for the class data. For the example above, the standard deviation for males was 1.73 hours, and the standard deviation for females was 2.05 hours.
- Next, calculate a *standard error* by dividing the standard deviation by the square root of the number of samples. For example, the researcher above surveyed 10 males and 10 females. So she divided the standard deviation for males (1.73) by the square root of 10. The standard error for males was 0.55. Using the same method (dividing the standard deviation, 2.05, by the square root of 10), she found that the female standard error was 0.65.
- The 95% confidence interval is the average ± 1.96Xstandard error. In other words, multiply the standard error by 1.96. Then add this number to the average to get the upper limit of the confidence interval. Subtract it from the average to get the lower limit of the confidence interval. For example, for the male data above, multiply the standard error (.55) by 1.96 to get approximately 1.1. Then add this number to the average for males (4.2 hours) to get the upper limit of 5.3 hours. Subtract 1.1 from the average to get the lower limit of 3.1 hours.

10. Review your hypothesis for question 3. On a separate sheet of graph paper, draw a graph to help you organize and interpret the data. The y-axis should indicate the average number of trips taken by students on public transportation. The x-axis will vary according to your hypothesis. Calculate 95% confidence intervals using the formula above, and add these to your graph.

DISCUSSION

11. Did the data support your hypothesis for question 1? Why or why not?

12. Why do you think students at your school made the transportation choices shown by your graph? Consider the local roads, public transportation system, the distance between student homes and campus, or anything else that might be relevant.

13. Can you use your class's survey results to generalize about the habits of all students around the country? Of non-students in your area? Why or why not?

14. Compare the results from your school to the US Census results for working Americans in 2000. Do the students at your campus differ from the census results? If so, why do you think that is the case?

15. Did the data support your hypothesis for question 2? Why or why not?

16. Did the data support your hypothesis for question 3? Why or why not?

17. Transportation choices can affect the environment greatly. Pick two different modes of transportation and explain how each one affects the environment. You may need to do some outside research to answer this question.

18. In your opinion, what factors are most important in determining how students choose between transportation options?

19. If you were to design another survey of student transportation choices, what would you change? Why?

TRANSPORTATION SURVEY

FIRST ASK: Have you already answered a transportation survey today? If so, stop the survey and move on to ask other students.

1. What mode of transportation do you use most often to get to school? Circle one:

   Drive alone   drive in a carpool   walk

   public transportation bicycle   other

2. How many miles away from school do you live? Circle one.

   Less than 5   between 5 and 10   more than 10

3. How many times did you use public transportation in the past 30 days?

4. Please fill out the following information about yourself:

   a. Are you male or female?

   b. Are you a full-time student or a part-time student?

SPREADSHEET HINTS

There are many ways to organize a spreadsheet for this lab. Following is one example that allows you to manipulate the data easily.

	mode of transportation (type 1 in the appropriate column)						distance from school (type 1 for less than 5 miles, 2 for between 5 & 10 miles, and 3 for more than 10 miles)	public transport	gender (F=female, M=male)	student status (F=full, P=part)
Respondent	drive	carpool	walk	p transport	bicycle	other		# trips		
1										
2										
3										
4										
5										
6										
7										
8										
9										
10										

Once all students have entered each respondent on a new row, use the spreadsheet to sum the number of students driving, carpooling, walking, etc at the bottom of each of those columns. For example, in Microsoft Excel, do this by typing =sum(data range). If the data is in column B, rows 3–150, type =sum(B3:B150). This is the data students need to make the first graph.

Next, make 3 copies of the entire spreadsheet. Leave the original spreadsheet as a reference. Use one of the copies for data sorted by distance from school, another for data sorted by gender, and the last for data sorted by student status. Students will need these data for their second and third graphs.

To obtain the necessary data for distance from school, follow these steps.

Highlight the entire dataset, and use the sort function to sort by distance from school. Next, insert at least 5 rows between the data for students less than 5 miles from school (coded as 1), between 5 and 10 miles from school (coded as 2) and more than 10 miles from school (coded as 3). These rows will give you some room for computing averages, standard deviations, standard errors, and confidence intervals for each set of students.

For students living less than 5 miles from school, use the sum function to add up the number of students driving, carpooling, etc, in each column at the bottom of this section. Repeat for students 5-10 miles from school, and for students more than 10 miles from school. These are the data needed for the second graph.

For the third graph, you will need to calculate each of these: the average number of trips on public transportation, the standard deviation, the standard error, and the confidence intervals. You will do each of these calculations for each sorted group of students (living less than 5 miles from school, between 5 and 10 miles from school, and more than 10 miles from school). For example, if you had data for students living less than 5 miles from school in column I, rows 3–50, you would use the following formulas in Microsoft Excel: for the average type =average(I3:I50); for the standard deviation type =stdev(I3:I50); for the standard error type =stdev(I3:I50)/sqrt(47) (47 because in this example there were 47 student respondents in this category); for the 95% confidence interval type =1.96*(stdev(I3:I50)/sqrt(47)). Add the 95% confidence interval to the average to get the upper limit, and subract it to get the lower limit. These are the data students will need for their third graph.

Repeat the steps above for gender and student status.

E X E R C I S E 7

Using Scatter Plots to Observe Patterns and Generate Hypotheses

OBJECTIVES

Students who complete this exercise will
- conduct surveys of student environmental views and habits
- plot three graphs to illustrate survey results
- identify correlations and create hypotheses about the causes of correlations
- design experiments to test their hypotheses

INTRODUCTION

Scientists draw graphs for many purposes. Sometimes, graphs help analyze data and evaluate hypotheses. At other times, scientists use graphs to visualize data and to help generate ideas for new hypotheses. In this exercise, you will gather data, graph it, and use it to write testable hypotheses that could explain patterns in the data.

Scatter plots look for relationships (*correlations)* between two or more variables. For example, in the first panel below, whenever one variable increases, the other also increases. In other words, the two variables are *positively correlated*. In the second panel, when variable 1 increases, variable 2 decreases. The two variables are *negatively correlated*. In the last panel, there is no obvious relationship between the two variables - they are *uncorrelated*.

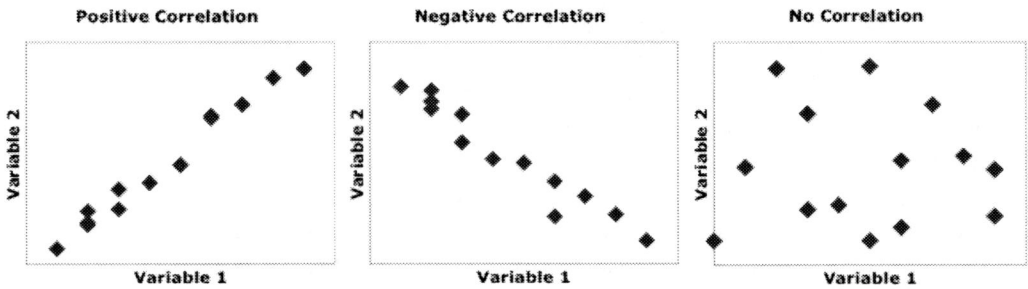

When environmental scientists find correlations, they can then hypothesize about why the two variables seem to be related. For example, if the concentration of a particular air pollutant is positively correlated with the incidence of asthma, they might hypothesize that the air pollutant leads to asthma, and design experiments to test this hypothesis.

It is important to remember that when two variables are correlated, that does NOT always mean that one of the variables caused the other. Although one variable *may* have affected the other, it's also possible that a third variable influenced both original variables. To determine the true causes of a correlation, a scientist has to conduct a controlled experiment.

For example, imagine a researcher wanted explore relationships between TV watching and exam grades. The researcher polled a group of students about how much TV they watched and their exam grades, then created the following scatter plot.

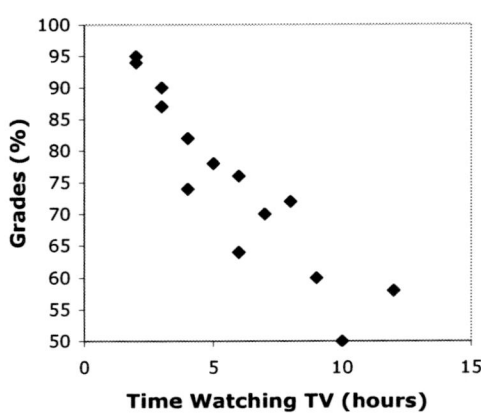

As the graph shows, watching TV and grades are negatively correlated. In other words, when TV watching increased, grades decreased. But this does NOT necessarily mean that watching TV caused low grades. It's also possible that students who spent more time watching TV had fewer hours to spend studying, and fewer hours of studying actually caused the lower grades. In this scenario, if a student had enough time to watch lots of TV *and* to study, they probably would have earned a high exam grade.

To determine if TV watching really causes low grades, the researcher would have to conduct a controlled experiment testing the hypothesis, "Students who spend more time watching TV will earn lower exam grades." To test this hypothesis, he could recruit a large group of students to take a test. He could randomly assign some students to watch no TV (a control treatment), others to watch a few hours of TV, and others to watch many hours of TV. He would make certain all students spent exactly the same time studying (time studying would be a control variable). If subjects who spent the most time watching TV in this experiment scored worse than those that watched little TV or no TV, that would support his hypothesis. If TV watching had no effect on grades, the researcher would need to reject his hypothesis and look for another explanation for the correlation. As you can see, although correlations don't explain what causes relationships, they do often give us ideas for hypotheses that can later be tested experimentally.

During today's laboratory exercise your lab group will survey students on your campus to explore whether their experiences, attitudes, and behaviors about the environment are related. You will use scatter plots to search for correlations, then write testable hypotheses about your observations.

MATERIALS

- At least 3 sheets of graphing paper per student
- 5–10 copies of "Environmental Biology Survey" per student
- 1 pencil or pen and 1 clipboard per student

METHODS

Each group of 3–4 students will conduct and analyze their own survey. You and your group will spread out around campus to survey students. Consider surveying students in the cafeteria, library, outdoor areas, study lounges, and hallways. Each group member should ask between five and ten students to fill out the environmental biology survey (your instructor will determine the exact number).

Before conducting surveys, your group needs to make a plan to avoid sampling bias. For example, how will you ensure that no respondents take more than one of your group's surveys? How will you ensure that your survey includes a representative sample of students (a group of students that resembles the student body in age, major, ethnicity, and other characteristics)? You may consider assigning certain group members to sample in different areas or to look for different groups of students. How will you ensure that you and your group members don't influence the answers respondents give you?

1. Outline your group's plan for conducting the survey in a way that minimizes sampling bias.

Once your instructor approves your plan, go ahead and conduct the survey. After you've collected enough student responses, return to the laboratory. Pool your data with the rest of your group.

RESULTS

To visualize the data you've collected, you will plot a series of scatter plots on separate sheets of graph paper.

2. Plot three graphs, each depicting the relationship between the answers to two survey questions. You can pick any three combinations of questions to depict. Make sure that in each graph you label the axes, include units where appropriate, and give your graph a title. Also make good use of space by setting scales so that most of the space of the graph is taken up with data points.

DISCUSSION

Interpret each graph that you plotted by answering the questions below.

3. Interpretation of first graph

    a. What variables did you graph?

    b. Were the variables correlated? If so, were they positively or negatively correlated?

    c. Why do you think there was a relationship (or lack of a relationship) between these variables?

    d. How would you design an experiment to test whether one of these variables causes another? Be sure to include in your explanation a hypothesis, a control treatment, and explain what results would support your hypothesis.

4. Interpretation of second graph

    a. What variables did you examine?

    b. Were the variables correlated? If so, were they positively or negatively correlated?

    c. Why do you think there was a relationship (or lack of a relationship) between these variables?

    d. How would you design an experiment to test whether one of these variables causes another? Be sure to include in your explanation a hypothesis, a control treatment, and explain what results would support your hypothesis.

5. Interpretation of third graph

    a. What variables did you examine?

    b. Were the variables correlated? If so, were they positively or negatively correlated?

    c. Why do you think there was a relationship (or lack of a relationship) between these variables?

    d. How would you design an experiment to test whether one of these variables causes another? Be sure to include in your explanation a hypothesis, a control treatment, and explain what results would support your hypothesis.

6. How would you improve this survey?

7. If you were designing a survey to study student attitudes about the environment, what would you ask? Suggest one additional question.

ENVIRONMENTAL BIOLOGY SURVEY

1. How many times in the last month have you walked in a park, biked, fished, canoed, picnicked, gone swimming outdoors, camped, or participated in other outdoor activities?

2. Approximately how many times did you go on overnight camping trips before you turned 18?

3. On a scale of 0 to 5, how important are environmental issues to your voting decisions? Please circle one number.

0	1	2	3	4	5
not important		somewhat important but other issues are more critical			extremely important

4. Approximately what percentage of the paper you discard do you recycle?

5. How many times in the past year have you visited a farmer's market?

6. How many courses have you taken in high school or college where some class time was spent discussing environmental issues?

E X E R C I S E 8

Competition in Plants

OBJECTIVES

Students who complete this exercise will
- measure growth and/or germination of plants grown at varying densities
- use the scientific method
- design and conduct experiments to test hypotheses about intraspecific and interspecific competition between plants
- illustrate their results with graphs

INTRODUCTION

Competition occurs when two organisms struggle for the same *limiting resource*. A limiting resource is a resource such as food, space, water, or nutrients that both competitors need, but that is in short supply. Both organisms expend energy and effort to use the resource before their competitor uses it. For example, a plant might grow longer roots so it can reach scarce water before neighboring plants. An animal might spend time patrolling and defending a territory and its food sources. As a result of the extra effort spent competing, competitors have less energy available for growth and reproduction. Therefore, competition harms each competitor, often by reducing growth and fecundity.

When members of a single species compete, it is called *intraspecific competition*. When competition occurs between organisms from different species, it is called *interspecific competition*.

Plants often compete for space, sunlight, water, nutrients, and minerals. When plant populations are dense, with many plants crowded together, resources run out more quickly and competition is more intense. As a result of competition in these crowded conditions, plants may grow more slowly, make fewer seeds, and those seeds may have lower germination rates.

Plants have evolved adaptations to help them compete. For example, some plants make large seeds with lots of stored food to help seedlings grow fast, so they can grow tall before their competitors can shade them. Others have wide leaves that cast large shadows on competitors. And others have root systems that are longer or wider than those of their competitors.

Today, you will set up an experiment to look for competition between plants. Specifically, you will try to answer this question: At what densities do radish and wheat seedlings experience interspecific and/or intraspecific competition?

OBSERVATIONS

Inspect the radish seeds, radish seedlings, wheat seeds, and wheat seedlings. Notice the seed sizes, leaf shapes, plant heights, root lengths, and anything else you think may be important.

1. Which species do you think will be a better competitor? What characteristics do you think will give it the edge?

At low density (few plants per pot) competition may be light, because resources won't be limiting. But at high density (many plants per pot), plants will start to crowd each other. They may shade each other, or they may run out of water, nutrients, or minerals. Look carefully at the size of the seedlings and the size of the pots.

2. At what density (the number of plants per pot) do you think competition will become important between radishes grown together in one pot?

3. At what density do you think competition will become important between wheat grown together in one pot?

HYPOTHESES

In this experiment, you will plant some pots with only radishes, others with only wheat, and others with both species together. For each combination (radishes only, wheat only, and both species) you will plant a variety of densities. After 3–4 weeks, you will measure growth or germination of the plants.

4. Write a hypothesis predicting the density (the number of plants per pot) at which intraspecific competition will decrease growth or germination for radishes.

5. Write a hypothesis predicting the density at which intraspecific competition will decrease growth or germination for wheat.

6. Write a hypothesis predicting the density at which interspecific competition will decrease growth or germination for radishes or wheat. Which plant do you expect to show the greatest decrease in growth or germination?

MATERIALS (PER GROUP)

- Wheat seeds
- Radish seeds
- Soil
- 15–30 3 inch plant pots
- 1 Scale
- 1 Ruler
- 1 Trowel
- A few radish seedlings and wheat seedlings, 3–4 weeks old, for the entire class to observe

METHODS

To detect competition between plants, scientists often use the following basic steps which your group will need to modify to test your own hypothesis.

FOR DETECTING INTRASPECIFIC COMPETITION

- Plant seeds at three or more densities, between 1 and 30 seeds per pot, as appropriate for you hypothesis.
- Place the pots in a sunny, warm spot for 3–4 weeks and water regularly.
- Measure growth and/or germination, as appropriate for your hypothesis.

FOR DETECTING INTERSPECIFIC COMPETITION

- Plant each species alone in various densities. (Use the pots you planted to test for intraspecific competition.)
- Plant the two species together. The total density of these treatments should equal the density of the single-species treatments. For example, if you planted 10 radishes alone, and 10 wheat alone, your treatment with both plants should have a total of 10 plants (5 radish and 5 wheat).
- Place the pots in a sunny, warm spot for 3–4 weeks and water regularly.
- Measure growth and/or germination, as appropriate for your hypothesis.

7. There are two independent variables (factors that you will change on purpose) in this experiment. One is density – you will have pots with varying densities of plants. The other is species – different pots will have different species planted. For each independent variable below, explain the *treatment levels* you will use:

   Density (the number of seeds per pot) treatment levels:

   Species (type of species per pot) treatment levels:

8. Name several *control variables* in this experiment.

9. What dependent variable will you use? In other words, what exactly will you measure when your plants have grown? How did you choose this measurement?

10. How many replicates will you have? Explain what your replicates will be.

11. Briefly explain how you will conduct your experiment. Make sure you give enough details that a reader could repeat your experiment. If necessary, draw a simple diagram to illustrate your experimental methods.

12. On a separate sheet of paper, make a chart for recording your data.

Discuss your experiment with the instructor before you continue. Once the instructor approves your plans, go ahead and begin the experiment.

RESULTS

13. Record the results on your data table.

14. On a separate sheet of graph paper, plot a graph depicting the results. Make sure you place the dependent variable on the y-axis (the vertical line). On the x-axis plot the density. Use different colors for different species treatments. Make sure you label the axes and put an appropriate scale on your axes (remember, you want the data to take up most of the space on the graph).

DISCUSSION

15. Were your hypotheses supported or rejected? For each one, back up your answer by referring to the data.

    Intraspecific competition in radishes

    Intraspecific competition in wheat

    Interspecific competition between radishes and wheat (make sure you compare treatments with both species to controls with one species)

16. Did either plant benefit from interactions with its neighbors? If so, at what densities? Explain by referring to your data.

17. If you had to do this experiment again, how would you improve it? You can always improve every experiment.

EXERCISE 9

Energy Conservation

OBJECTIVES

Students who complete this exercise will
- calculate and compare the costs of efficient and inefficient light bulbs
- measure power use and calculate energy costs of using computers
- measure and calculate the costs of "vampire energy" used by campus appliances
- discuss the rewards and difficulties of energy conservation

IINTRODUCTION/OBSERVATIONS

Residents of the United States have the highest per capita consumptive energy use in the world. We use commercial energy primarily for transportation, heating and cooling, and for electricity. Our high energy use leads to many problems, including expensive utility bills, dependence on other nations, air pollution, and depletion of resources. We could ease all of these problems if we conserved more energy.

One way to conserve energy is to change our behaviors. For example, we could drive less, walk more, and take public transportation. We could turn off lights when we don't need them, and lower thermostats in the winter.

Another way to conserve energy is to use more efficient machines. *Energy efficiency* is the percentage of energy used in a machine that actually performs useful work. For the most part, our machines have low energy efficiency, and the vast majority of commercial energy that we buy is wasted. For instance, most of the energy in the gasoline we pump into a car does not actually move the car forward. Instead, most of it creates waste heat that has to be blown away by a fan to prevent the car from overheating.

Why is most of the energy wasted? First, as the second law of thermodynamics states, when you convert energy from one form to another (for example, from the potential energy in gasoline to the kinetic energy of a car moving forward), some of the energy is always converted to heat or another type of high-entropy energy. This is a physical law that we cannot overcome—machines can never be 100% efficient. Second, manufacturers usually don't make machines that are as efficient as technology allows, because often this makes the machine more expensive.

Environmental scientists measure two different types of costs. *Initial cost* is simply the purchase price you pay for an item at the store. *Life cycle cost* is the initial cost *plus* the price of using and maintaining that item for as long as you own it. For example, the life cycle costs of a car include the sticker price plus the amount you spend on gasoline, oil changes and other maintenance.

Often, highly efficient machines have a higher initial cost, but a lower life cycle cost than comparable low efficiency machines.

In this laboratory exercise you will investigate the energy efficiency and costs of machines on campus. Specifically, you will
- calculate costs of using incandescent versus fluorescent light bulbs
- measure energy use of computers that are actively running, using screen savers, and asleep
- measure energy use of appliances or machines around campus that are turned off

MATERIALS (SHARED BY THE CLASS)

- 2 identical lamps
- 2 light bulbs that create equivalent light brightness (measured in lumens), one incandescent and one Energy Star rated compact fluorescent, with original packaging showing the price tags of each
- 1 recent utility bill for the campus or for a nearby residence, showing the price per kilowatt hour charged by the local utility company

ADDITIONAL MATERIALS (PER GROUP)

- 1 computer (alternatively, the whole class can share one computer)
- 1 Kill a Watt meter
- 1 calculator

PART 1: COMPARING THE ENERGY EFFICIENCY OF LIGHT BULBS

Your instructor has purchased 2 light bulbs that are equally bright; one is a traditional incandescent light bulb and the other a compact fluorescent light bulb. At the beginning of lab, your instructor will place these bulbs in identical lamps and turn them on. Examine the price tags on the two types of light bulbs.

1. What is the initial cost of each light bulb?

   Incandescent initial cost = _____

   Fluorescent initial cost = _____

2. Which light bulb has a higher initial cost, and by how much?

3. What is the wattage of each light bulb as indicated on the packaging? This is a measure of how much power it uses when lit (and therefore is also related to how much energy you need to light it).

    Incandescent # of watts = _____

    Fluorescent # of watts = _____

4. Say you turn one of the lights on for 10 hours a day, every day of the year (365 days). How many hours is it lit during a year?

    10 hours X 365 = _____ hours used in one year

5. For each light bulb, how many kilowatt hours will it use in a year?

    Incandescent:

    (_____ X _____)/1000 = _____ kilowatt hours per year
     # Watts   hours used in
                        one year

    Fluorescent:

    (_____ X _____)/1000 = _____ kilowatt hours per year
     # Watts   hours used in
                        one year

6. Examine the local utility bill provided by your instructor. How much does the local utility charge per kilowatt hour?

7. How much will it cost to run the light bulbs for one year?

    Incandescent:

    _____ X _____ = _____ operating cost
    Cost/kilowatt hour   kilowatt hours per year

    Fluorescent:

    _____ X _____ = _____ operating cost
    Cost/kilowatt hour   kilowatt hours per year

8. Your total costs over the first year:

    Incandescent: _____ + _____ =
                  Initial cost   operating cost

    Fluorescent: _____ + _____ =
                 Initial cost   operating cost

9. Which light bulb has a higher life cycle cost?

10. Estimate how many light bulbs you have in your home. If each one was incandescent (and you burned them each for 10 hours a day), how much would you spend on them during the first year?

11. If each of your bulbs at home was fluorescent (and you burned them each for 10 hours a day), how much would you spend on them during the first year?

12. How much would you save on lighting costs in your entire house by using compact fluorescent bulbs for one year, compared to incandescent bulbs?

13. Besides saving money, what other benefits do you (or society) get from using energy more efficiently? List and explain at least three.

14. Go to the front of the class and put your hand underneath each light bulb (but DON'T touch the bulb!). Which is wasting more energy as heat?

PART II: ENERGY USE BY COMPUTERS

The amount of energy used by a computer varies according to what the computer is doing. Plug your Kill A Watt meter into an outlet, and plug a computer into the meter. Make sure the meter is set to measuring Watts.

15. How many watts does the computer use during each of these activities?

      Computer turned off _____

      Computer actively running 1 or 2 programs _____

      Computer running a screen saver _____

      Computer asleep _____

16. If the school left the computer asleep for 12 hours a night, every night for a year, how much would the electricity cost? To calculate this, follow the steps you used for the light bulbs. First, calculate hours used in one year, then the kilowatt hours per year, then multiply by the local utility rate to find the operating cost. Show your work.

17. If the school didn't put computers to sleep, but instead let them run a screen saver for 12 hours a night for a year, how much more would the electricity cost? Show your work.

18. How much would the school save by turning computers off instead of putting them to sleep?

PART III: VAMPIRE ENERGY

Many machines use a little bit of energy even when they are turned off. This energy is often called "vampire energy." For example, most televisions use some energy whenever they are plugged-in, even if nobody is watching anything. Stereos, microwaves, copy machines, and other machines often do this as well.

With your group, find a machine on campus that is turned off, yet drawing some power. Make certain you have permission to unplug the machine before you do it. Your instructor may have guidelines about this. Measure how many watts the machine is using while turned off.

19. What machine did you measure, and how many watts was it using? If the machine you measured did not draw vampire energy, try another machine until you find one that does draw vampire energy.

20. How much would the school pay for electricity per year if this machine was plugged in 24 hours per day for a year, but not turned on? Show your calculations.

ADDITIONAL QUESTIONS

Although conserving energy has many benefits for society, it can be difficult to achieve. Habits, finances, politics, and even psychology can put up barriers to conservation.

21. In your own life, name several behavioral changes that you could make to save energy.

22. Which of the behavioral changes you listed would be easiest to achieve? Why?

23. Which would be hardest to achieve? What difficulties would you encounter if you tried to make this change?

24. Consumers often face a dilemma. They can save money up front by purchasing a machine with a lower initial cost (but higher life cycle cost), or save even more money in the long run by purchasing a machine with a lower life cycle cost (but higher initial cost). Suggest three steps that society could take to encourage consumers to purchase more efficient machines.

E X E R C I S E

Transpiration

10

OBJECTIVES

Students who complete this exercise will
- measure transpiration rates of campus plants
- use the scientific method
- design and conduct experiments to test hypotheses about transpiration rates
- illustrate their results with graphs

INTRODUCTION

Water is essential for life on Earth. One reason organisms need water is that it facilitates thousands of chemical reactions in cells by dissolving the molecules involved. Without water, these chemical reactions simply don't occur and cells die. Water also helps regulate temperature and maintain pH. In plants, water pressure helps support non-woody plants (and therefore plants wilt when they lack water) and it also acts as a key ingredient in photosynthesis.

A variety of biological and physical processes move water from the atmosphere to Earth's surface and back to the atmosphere in a phenomenon called the *hydrologic cycle*. Plants play a major role in the hydrologic cycle, because collectively all the plants on Earth move enormous amounts of water from one place to another. For example, plants pull water out of the ground with their roots and move it above ground to leaves and other tissues. From the leaves, liquid water escapes back into the atmosphere as water vapor. Eventually, this water vapor condenses into clouds and falls back to the ground as precipitation.

Transpiration occurs during photosynthesis. During photosynthesis, plants take carbon dioxide from the air and use it to make sugars. To obtain the carbon dioxide, leaves have small pores called *stomata,* which they can open and close. When the stomata are open, carbon dioxide enters cells and plants can photosynthesize. However, although the stomata are open to let carbon dioxide enter, water can exit leaves as water vapor in the process called transpiration.

Transpiration rates depend on many factors. For example, since water is lost through stomata, anything that affects how many stomata are open will also affect the rate of transpiration. Some plant species have more stomata per unit area of leaf, and therefore can have higher transpiration rates than other species. When plants have abundant light available, they tend to open stomata because they need more carbon dioxide for photosynthesis. Therefore, higher light intensity can increase transpiration rates. When water is scarce, plants close their stomata to preserve water, decreasing transpiration rates. Transpiration rates also depend on humidity, temperature, and wind. Low humidity and high wind both facilitate water loss from stomata.

Today, you will observe transpiration and design an experiment to test how one factor affects transpiration.

OBSERVATIONS

To help you formulate a hypothesis, first, examine plants that are likely to photosynthesize.

1. Look for as many of the following plant pairings as you can find. For each one, make a prediction. Which one will have higher transpiration rates? Then, for each one, explain how you decided on your prediction.

    a. Leaves in the shade vs. leaves in bright sunlight

    b. Young leaves vs. old leaves

    c. Evergreen leaves vs. deciduous leaves

    d. Tree leaves vs. herbaceous plant leaves

2. Describe another pairing of leaves that you observed (besides the ones listed above) that might have different transpiration rates. Which one do you predict will have higher transpiration rates? Why?

HYPOTHESIS

Today you will test one of the predictions you made above. You will place plastic bags around leaves to catch the water exiting leaves through transpiration. Then, you will weigh the bags and use the weights to calculate rough estimates of transpiration rates for each type of leaf.

3. Pick one of the predictions you made in questions 1 or 2, and use it to write a testable, precise hypothesis for your experiment today.

MATERIALS (PER GROUP)

- 10–15 clear gallon plastic bags (not zipper bags)
- 10–15 twist ties
- Scale that can accurately weigh 0.1g or less
- Tape and labeling pen
- Paper towels or cloth towels

METHODS

Estimate transpiration rates using the following basic steps which your group will need to modify to test your own hypothesis. Read the instructions below, then plan your experiment with your group using the questions below as guidelines.

MEASURING TRANSPIRATION

- Begin by carefully weighing each bag ahead of time, in grams. If all bags are identical, your instructor might allow you to weigh just one bag and assume the rest are the same—but be careful about this because small differences in weight will be a problem for this experiment. Record each bag's weight on labeling tape along with your team's initials. Attach the tape to the end of a twist tie, folding it in half. Twist the tie around the bag.
- Gently place a bag over a group of leaves, enclosing them completely. Try to find leafy stems without flowers or berries for this experiment. It is also important that all the leaves are dry before beginning this experiment. Wrap the end of the bag around the twig and close it tightly with the twist tie.
- The label already lists the bag weight and the group initials. Add the exact time when the bag was sealed over the leaves to the label, as well as the type of treatment (for example, sun or shade). Make sure the label does not block sunlight from reaching the leaves inside.

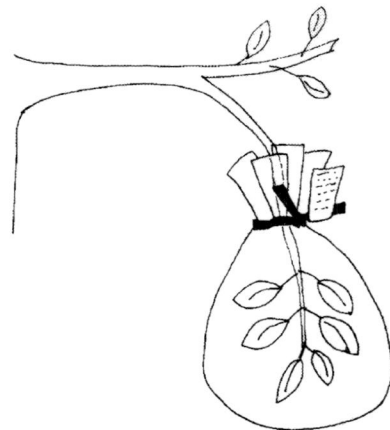

- Continue placing bags around all of your replicates.
- Exactly 1 hour after you sealed each bag, collect the bag and leaves as follows. First, shake the leaves and bag so that water runs into the bottom of the bag (it will not all go to the bottom – that's ok). Untwist the twist tie and gently remove the bag, along with all of the leaves that it enclosed, making sure to keep all the water in the bag. You will have to pull the leaves from the plant as you do this. Do not collect twigs or other non-leaf parts that were in the bag. Quickly retwist the bag shut near the top so that water does not leak or evaporate out.
- After you collect all samples, weigh them carefully as follows.
- Weigh the bag with the water and leaves inside, but without the twist tie and label.
- Remove the leaves and pat each one gently dry with a paper towel or cloth towel to remove all water from its surfaces. Weigh the dry leaves alone.
- For a rough estimate of transpiration rate, use the following calculations, making sure all weights are measured in grams:

  (Weight of bag + water + leaves) – (weight of bag alone) - (weight of leaves alone) = weight of water alone

  (weight of water alone) / (weight of leaves alone) = grams water released / grams leaf per hour

4. The *independent variable* in this experiment will depend on the hypothesis you decided to test. What is your independent variable?

5. What are the treatment levels of your independent variable? For example, you might have a shade treatment and a light treatment, or an evergreen treatment and a deciduous treatment, etc., depending on your hypothesis.

6. You should also have a *control treatment* for this experiment. This treatment will check to make sure you are really measuring transpiration (water leaving a leaf) rather than simply water from twigs or the atmosphere condensing on the bags. Think carefully – what kind of control treatment could you use that would allow you to distinguish between transpiration and simple condensation? Explain in detail.

7. Name several *control variables* in this experiment.

8. The *dependent variable* will be what you measure. What exactly will you measure for each treatment?

9. How many *replicates* will you have? Explain what your replicates will be.

10. Briefly explain how you will conduct your experiment. Make sure you give enough details that a reader could repeat your experiment. If necessary, draw a simple diagram to illustrate your experimental methods.

11. On a separate piece of paper, make a chart for recording your data.

Discuss your experimental design with the instructor before you continue. Once the instructor approves, carry out the experiment.

RESULTS

12. Record the data on your data chart.

13. On a separate sheet of paper, plot a graph depicting the results. Make sure you place the dependent variable (grams of water released per gram of leaf per hour) on the y-axis (the vertical line). On the x-axis plot the independent variable. Make sure you label the axes and put an appropriate scale on your axes (remember, you want the data to take up most of the space on the graph).

DISCUSSION

14. Was your hypothesis supported or rejected? Back up your answer by referring to your data.

15. For each treatment of your independent variable (for example for light and for shade) calculate the average number of grams of water released by each gram of leaves per hour. Then, calculate how much water per gram of leaf would be lost to transpiration during one day if these leaves transpired at the average rate for *12 hours* during the day. Show your work.

16. If the plant had 1000 grams of leaves, how much water would it the plant lose if it tran-spired at the average rate for 12 hours? Calculate this for each treatment, and show your work.

17. How realistic are your calculations for questions 13 and 14? Does the plant really tran-spire at the same rate all day? Give a detailed explanation of why you would or would not expect this.

18. For each treatment, how much water would be lost to transpiration during one year if the plant had 1000 grams of leaves and transpired at the average rate for 12 hours every day, all year? Show your calculations.

19. How realistic is your calculation for question 16? Does the plant really transpire at the same rate every day during the year? Give a detailed explanation of why you would or would not expect this.

20. What happens to water after it is transpired?

21. If you had to do this experiment again, how would you improve it? You can always improve every experiment.

E X E R C I S E *11*

Soil Invertebrate Diversity

OBJECTIVES

Students who complete this exercise will
- collect and identify soil invertebrates
- calculate Simpson's Diversity Index
- use the scientific method
- design and conduct experiments to test hypotheses about soil invertebrate diversity
- illustrate their results with graphs

INTRODUCTION

Biodiversity is the variety of life on Earth; it is among the most fascinating aspects of nature. Millions of species and life forms have evolved over Earth's history, including birds, insects, fishes, corals, trees, fungi, and more. Ecologists seek to understand why some ecosystems are more diverse than others. They also struggle to predict how the current worldwide loss of biodiversity will affect ecosystem services in the future.

There are many ways to measure biodiversity. When studying species diversity, scientists often measure *species richness*, the number of species in an ecosystem. Species richness is a simple measure of diversity, but unfortunately it is greatly influenced by sampling effort. For example, say you measured the species richness of birds in a meadow by observing the meadow carefully for three hours and recording each type of bird you saw. If you observed 4 different bird species during those three hours, then species richness for birds in that meadow is simply 4 species. However, if another student observed the same meadow for an entire week, that student might see more types of birds, and report higher species richness.

Another measure of species diversity is *evenness*. Evenness describes the proportions of individuals of different species in an ecosystem. When proportions of individuals are similar, an ecosystem has high evenness. For example, say you watched the meadow and counted 100 individual birds. If 97 of those individual birds were crows, and you also saw one sparrow, one blue jay, and one cardinal, then the meadow had low evenness. In other words, the vast majority of birds in the meadow belong to one species, crows, and thus diversity is low. If instead you saw 25 crows, 25 sparrows, 25 blue jays, and 25 cardinals, then the meadow has high evenness, because it has similar numbers of all species.

A variety of diversity indexes calculate species diversity using equations that account for both species richness and evenness, and thus give more complete measures of diversity. In this laboratory exercise, you will calculate Simpson's Diversity Index, a common index used by ecologists. Simpson's Diversity Index is highest when species richness and evenness are both high.

For this laboratory you will compare the diversity of soil invertebrates in two locations. In particular, you will study invertebrates that commonly live underneath rocks or logs such as pill bugs, earthworms, springtails, and centipedes. Many of these animals feed on detritus, while others prey on the detritus feeders. Many are nocturnal, rarely emerging from underneath rocks or logs during daytime. Soil invertebrate diversity could be affected by many factors that differ from place to place, such as the availability of detritus or other food sources, soil pH, soil moisture, the vegetation type, temperature, the amount of sun exposure, and more. On the first day of this laboratory exercise, your group will choose two locations to compare; then design and set up your experiment. Approximately one week later, you will collect samples, identify the organisms, calculate diversity, and interpret your data.

OBSERVATIONS

To help you formulate a hypothesis, first, observe the study area, which could consist of lawns, forests, flowerbeds, fields, or other habitat types as determined by your instructor.

1. With your lab group, examine the conditions in a variety of locations within the study area, and answer the following questions.

    a. Are any locations particularly sunny? Which?

    b. Do any locations have more detritus than others? Which?

    c. Do any locations have higher soil moisture than others? For example, areas near streams or at the bottom of hills usually have more moisture than high areas.

    d. What vegetation types can you find in the study area? For example, are there lawns, forests, flowerbeds, or others?

    e. What other differences do you notice? Describe and explain at least one additional difference between locations in the study area.

Next, look for stones, logs, sticks, or other objects in the study area under which small invertebrates could hide. With your group, pick up some of these objects and look underneath for invertebrates.

2. Did you see any animals? In what location did you find the greatest variety of animals?

HYPOTHESIS

Pick two locations that you observed above. You will compare invertebrate diversity at these locations.

3. Write a testable, precise hypothesis predicting which location will show greater soil invertebrate diversity, if any.

4. Why did you make the prediction that you did?

MATERIALS (PER GROUP)

Week 1:
*   6–10 8 inch X 8 inch untreated wooden boards (not plywood or particle board)
*   1 gardening trowel
*   soil pH or moisture detectors (optional)
*   Duct tape or water proof labels
*   Permanent marker or pen

Week 2:
*   6–10 garbage bags
*   at least 3 plastic tubs, each at least 4 inches deep
*   at least 2 pairs of disposable gloves per student
*   1 gardening trowel
*   1 ruler
*   1 dissecting microscope
*   1 pair of forceps per student
*   3–4 bowls, approximately 500 mL – 1 L
*   1 plastic spoon per student
*   dichotomous keys or guides to local soil fauna

METHODS

Collect and identify soil invertebrates using the methods outlined below. Remember that your group will need to modify these methods to test your own hypothesis. Use the questions below to help you plan your experiment.

COLLECTING SOIL INVERTEBRATES

- To place labels on the wooden boards, put pieces of duct tape several inches long on clean, dry spots on the boards. Press firmly. Label them with your team initials, the date, and the appropriate locations.
- Place boards flat on the ground at your study locations.
- Leave the boards in place for approximately one week.
- To collect the animals underneath the wooden boards, wear disposable gloves. Gently lift each board, and quickly pick up any rapidly moving organisms. Place them in a garbage bag. Collect leaf litter or other loose organic matter from under the board, and add it to the garbage bag. Using a garden trowel and a ruler, collect the top inch (or top 2.5 cm) of soil from underneath the board, and add it to the garbage bag.
- Label the bags with your team initials, the date, and the appropriate location.

IDENTIFYING AND COUNTING SOIL INVERTEBRATES

- In the laboratory, dump the contents of a garbage bag into a tub.
- Wearing disposable gloves, carefully sort through the soil, humus, or other materials, picking out all organisms with forceps and placing them in bowls. Make certain you examine all sides of leaves and detritus. Examine soil in small batches by picking a spoonful at a time, dumping it into a bowl, and looking for organisms. When you have thoroughly examined each piece of detritus or each small batch of soil, put them back into the garbage bag so you don't accidentally examine them twice.
- Once you've collected all animals in bowls, identify each one using dichotomous keys or guides.
- Use a dissecting microscope to help identify small animals.
- If you can't identify a species after trying diligently and consulting with your instructor, simply call it unknown #1, unknown #2, etc.
- Count the number of individuals of each species in each sample.

5. The *independent variable* in this experiment will depend on the hypothesis you decided to test. What is your independent variable?

6. What are the treatment levels of your independent variable? For example, you might have forest and meadow treatments, or high detritus and low detritus treatments, etc., depending on your hypothesis.

7. Name several *control variables* in this experiment.

8. The *dependent variable* will be what you measure. What exactly will you measure for each treatment?

9. How many *replicates* will you have? Explain what your replicates will be.

10. Briefly explain how you will conduct your experiment. Make sure you give enough details that a reader could repeat your experiment. If necessary, draw a simple diagram to illustrate your experimental methods.

11. On a separate sheet of paper, draw a chart for recording the data. Remember that for each sample, you will need to record the number of individuals of each species.

Once the instructor approves your experimental design, go ahead and place your boards at the appropriate locations.

RESULTS

Record the results on your data table. Then, answer the questions below.

12. Which species were most abundant in each location?

For each replicate sample, calculate the Simpson's Diversity Index using the following equation.

$$\text{Simpson's Diversity Index} = \frac{N\,(N\text{-}1)}{\Sigma\, n\,(n\text{-}1)}$$

In this equation, "N" is the total number of individuals of all species in the sample. The number of individuals in a single species is "n," and "Σ" refers to the sum for all species. For example, say you found 3 species in your sample, 10 earthworms, 5 springtails, and 1 centipede. The total number of individuals, N, is 16 (10+5+1). The Simpson's Diversity Index would be calculated as follows.

$$\text{Simpson's Diversity Index} = \frac{16\,(16-1)}{(10\,(10-1)) + (5\,(5-1)) + (1\,(1\text{-}1))} = 2.2$$

13. Calculate Simpson's Diversity Index for each replicate sample. Show your work.

14. On a separate sheet of paper, draw a graph comparing the diversity index at the different locations. Be sure to label your axes and set your scales so that you use up most of the space on the graph.

DISCUSSION

15. Was your hypothesis supported or rejected? Back up your answer by referring to the data.

16. Why do you think you got the results that you did?

17. How would you improve this experiment? Every experiment can be improved.

E X E R C I S E

Community Structure of a Forest

12

OBJECTIVES

Students who complete this exercise will
- walk through a forest and describe its community structure
- identify and describe forest layers
- find and describe examples of light competition among trees
- find and describe examples of predation, mutualism, parasitism, and commensalism in the forest

INTRODUCTION

Community structure is a comprehensive description of what a community is like, including descriptions of the organisms present as well as their abundance, diversity, distribution, and species interactions.

The class will walk through a forest to observe some facets of its community structure. As you walk, work with your group to answer the questions below. Every community is unique, making it impossible to predict what you will find today. You may not find everything listed below, and you will certainly see many additional organisms and interactions. Take notes as you encounter interesting organisms or interactions, and not necessarily in the order listed below. Have fun!

FOREST LAYERS

Most forests have layers of plants growing at different heights.

1. With your instructor's help, define each of the following layers. Then, using dichotomous keys, guides, or your instructor's guidance, name and briefly sketch a dominant plant in each forest layer.

   a. Canopy layer
      - Defininition

      - Dominant organism name and sketch

    b.   Understory layer
- Defininition

- Dominant organism name and sketch

    c.   Shrub layer
- Defininition

- Dominant organism name and sketch

    d.   Herb layer
- Defininition

- Dominant organism name and sketch

COMPETITION

Competition is a species interaction where two organisms struggle for a *limiting resource,* a scarce resource that is not present in sufficient quantities to meet the needs or wants of community members. One resource that is almost always limiting in a forest is sunlight.

   2. Compare the amount of light in a forest versus an open field. Which has more light? Why?

3. Which forest layer (canopy, understory, shrub, or herb) has the most amount of light available?

4. Which forest layer has the least amount of light available?

Plants have evolved a variety of adaptations for dealing with low light conditions and for effectively competing for light.

5. Search with your class for the following plant types. Briefly sketch the plants and for each one, describe an adaptation that helps it compete for sunlight.

   • vine

   • evergreen plant

The Competitive Exclusion Principle suggests that when species compete for the exact same resource over long periods of time, one is usually slightly better at attaining the resource. The more successful competitor will thrive and multiply, while other competitors will slowly decrease in abundance. Eventually, the better competitor will completely exclude inferior competitors from the community. The Competitive Exclusion Principle has been demonstrated primarily in laboratory settings and in simple habitats. But it doesn't always hold true. For example, the trees in the forest compete for light. If the Competitive Exclusion Principle held true, we would expect one tree species (the one best at competing for light) to eventually dominate the forest, excluding other species. Yet we see multiple tree species in a single forest.

6. Why doesn't one species of tree dominate and exclude all other species from this forest? There are many possible answers to this question. Discuss this with your group and the instructor, and then describe at least one reason why the Competitive Exclusion Principle doesn't apply in this forest.

7. Besides light, what else might be a limiting resource in a forest? Name at least several possibilities.

PREDATION

Predation is a species interaction where one species, the predator, benefits by eating another, the prey. With your class, search for tell-tale signs of predators in the forest. For example, you might look for spider webs, holes left in trees by woodpeckers, or birds of prey flying over the forest.

8. Name several predators and their prey that live in this forest.

Predators and prey have evolved in response to each other. Predators have evolved adaptations for finding and catching prey, while prey organisms have evolved adaptations for evading predators.

9. The class will search for examples of each of the following adaptations commonly seen in prey organisms. If you find them, draw a quick sketch or written description of the organism and its adaptation.

- Cryptic coloration/camouflage

- Chemical defense

- Other adaptations for evading predation

MUTUALISM

In *mutualism,* a close interaction between two species benefits both species. Search with your class for examples of mutualism.

10. A common mutualism in forests is pollination.
    a. How does a plant benefit from a visit by a pollinator?

    b. How does the pollinator benefit from a visit to a flower?

    c. Draw one flower that you saw during your trip to the forest. What adaptations does the flower use to attract pollinators?

11. Lichens are fungi and algae living together in mutualistic associations. Fungi provide nutrients and protection for the algae, while the algae provide food for fungi. Draw one lichen that you saw today.

12. List other examples of mutualism that you saw today.

PARASITISM

In *parasitism,* one species (the parasite) benefits from a close association with another (the host), while the host is harmed by the interaction. There are many parasites in a forest, ranging from deer ticks to mistletoe plants.

13. Search with your class for the following examples of parasitism. If you find them, draw a quick sketch. Explain for each one which species is the parasite, which is the host, and how the two species affect each other.

   • *plant galls*—growths caused by parasitic insects or bacteria on stems and branches

   • parasitic fungi on leaves

   • other examples of parasitism

COMMENSALISM

A species interaction where one species benefits from a close interaction with another, but the other is not helped or harmed by the interaction is called a *commensalism.*

1. Search with your class for the following examples of commensalism. If you find them, make a quick sketch. For each one, explain what species are involved, and how they affect each other.

    • Bird or squirrel nests on trees

    • Lichen or mosses on tree trunks

    • Other examples of commensalism

ADDITIONAL NOTES

If your instructor or a classmate pointed out any other organisms, interactions, or anything else you may want to remember about this forest, use this space to write additional notes for yourself.

SAFETY NOTE. Be sure to check for ticks after lab, as they are common in many forests and can carry diseases. If you see one, remove it completely. Familiarize yourself with the symptoms of Lyme disease, and see a doctor if they occur.

E X E R C I S E   13

Watersheds and Stream Ecology

OBJECTIVES

Students who complete this exercise will
- walk through watersheds and examine streams
- identify possible sources of pollution in watersheds
- identify impervious surfaces in watersheds
- find and describe riffles and pools
- collect and observe benthic organisms, nekton, and plankton

INTRODUCTION

The class will observe a watershed and its stream. As you walk, work with your group to answer the questions below. You may not find examples of everything listed below, and you will certainly see additional items of interest. Take notes as you go, and remember you will not necessarily find things in the order listed below. After the walk, you will take samples back to the laboratory to view in a microscope. Have fun!

MATERIALS (SHARED BY THE CLASS)

- 1 plankton net
- 1 seine net (optional)
- Dichotomous keys or guides for identifying local stream organisms (optional)

MATERIALS (PER GROUP)

- 1 bucket
- 1 ice-cube tray
- 1 dip net
- 1 compound microscope
- 5–10 slides
- 5–10 cover slips
- Protoslo ® (optional)

WATERSHEDS

Watersheds are land areas that drain into a particular stream or other body of water. Any precipitation that falls on a watershed could potentially work its way into the stream, either as runoff or through groundwater flow.

1. Name several landmarks, such as buildings, roads, hills, or trees in this watershed.

Land use in a watershed greatly affects the type and quantity of pollutants that enter rivers and streams. For example, roads can collect oil and litter. When it rains, the oil and litter often wash into storm drains and flow directly into streams. Farmers and homeowners sometimes add fertilizers and pesticides to their fields or yards, and these pollutants can also wash off in rainstorms and enter streams.

2. What potential pollution sources do you see in this watershed? Look for storm drains, roads, litter, areas that are treated with fertilizers or pesticides, or anything else that could lead to pollution.

Land use also affects how well watersheds filter and remove pollutants before they enter streams. In wetlands and forests, water infiltrates the ground and flows downstream slowly as groundwater flow. As groundwater trickles through the sediment, some pollutants stick to sediment particles. Soil bacteria break down other pollutants. Plants take up and remove fertilizers. In the end, clean water enters the stream.

Impermeable surfaces, such as roads and rooftops, prevent water from infiltrating into the ground. Instead, water flows across these surfaces as *runoff,* and often is shunted to storm drains that release the water directly into streams. As a result, more pollution enters streams.

3. Describe a place in the watershed with impermeable surfaces or surfaces with low permeability. Where does rainwater go when it hits this place? You may have to look for storm drains and gutters, examine the topography, or discuss this with your instructor.

In watersheds with a high percentage of impermeable surfaces, the volume of runoff builds quickly during a storm, leading water to enter the streambed in fast torrents. High velocity water erodes stream banks, eroding sediments into the stream. Erosion also exposes tree roots, making the trees along stream banks more vulnerable to falling over.

4. Do you see evidence of erosion along the stream banks? If so, describe the evidence.

STREAM ECOLOGY

Streams often have a variety of habitats that are home to different organisms. Two types of habitats often found in streams are *riffles* and *pools*.

5. With your instructor's guidance, find riffles and pools and describe these habitats in a few sentences.

    Riffles

    Pools

6. Which of the areas (riffles or pools) will most likely have the highest oxygen concentrations? Why?

7. Which will have the most detritus available for detritivores? Why?

We will try to observe some of the organisms that live in streams.

Benthic organisms live on or just above the streambed. Some benthic organisms live attached to rocks, some bury in sediments, and some crawl or swim along the streambed.

Benthic organisms include producers, such as algae attached to rocks or sediment.

   8.   Draw and describe a benthic producer that you saw today.

Benthic organisms also include many animals. For example, many flying insects such as dragon-flies or mayflies have larval stages that live as benthic organisms in streams. There are many good methods for finding and examining benthic animals. One method is to stretch a seine net across the stream, and gently rub rocks upstream from the net so that attached organisms are carried in the flow towards the net. Another method is to simply turn over rocks one by one, examining each for attached organisms. A third method is to gather detritus, such as fallen leaves, and place small clumps in buckets of water. Small benthic organisms often attach to the detritus and can be found by carefully examining each leaf. Your instructor will explain how your class will find benthic organisms. When you find an interesting organism, place it in a bucket or in an ice-cube tray with water, to show to your classmates later. When all students have seen all of the benthic organisms, gently return them to the stream.

   9.   Draw three benthic organisms found by your class. Use identification keys and guides to help you identify these organisms, if possible.

   10. Many benthic organisms have adaptations that help them hold on to rocks or mud to pre-vent being washed downstream in a storm. Did you see any adaptations among your ben-thic organisms to prevent being washed away? If so, describe one in detail.

Nekton are strong swimmers that can swim against currents. Fishes are common nekton in streams. Using dip nets, gently catch one or two fishes and place them into a bucket with water. Be sure to show your catch to the entire class. When everyone is done viewing the fishes, gently release them into the stream.

11.   Draw one type of nekton that you saw today. Use identification keys or guides to help you identify it, if possible.

Plankton are organisms that float. Although many planktonic organisms can swim to some extent, none can swim well enough to fight against a current. Therefore, plankton are carried from place to place by water movements. Your instructor will help the class collect plankton with a plankton net.

Be sure to wash your hands carefully and check for ticks when you are done with field investigations.

EXAMINING PLANKTON

When you return to the laboratory, view the plankton with a microscope. First draw a small ring of Protoslo ® (a compound that will slow down swimming plankton) approximately the circumference of a dime, near the center of a clean slide. Place a drop of stream water with plankton inside this ring. Cover the drop with a cover slip as shown below. Using a compound microscope, focus on the slide and scan the area for organisms. Your instructor will decide how many slides each group should make and view.

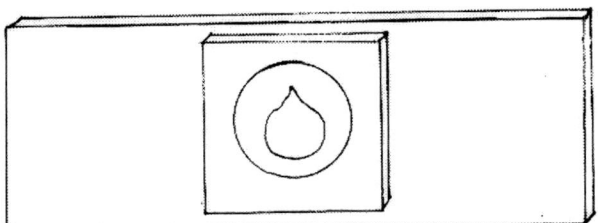

Phytoplankton are planktonic producers. The easiest way to spot phytoplankton is to search for the green or golden-brown pigments that help them photosynthesize. Phytoplankton are often very small, so look carefully. Once you've spotted these at lower magnifications, zoom in so you can draw them in detail.

12. Draw three phytoplankton that you found today. If identification keys or guides are available, try to identify your phytoplankton.

Zooplankton are planktonic consumers. Zooplankton eat phytoplankton, detritus, or other zooplankton. Zooplankton are often slightly larger than phytoplankton, and often swim quickly. They are frequently transparent.

13. Draw three zooplankton that you found today. If identification keys or guides are available, try to identify your zooplankton.